Environmental Dynamics of Pesticides

Environmental Science Research

Editorial Board

Alexander Hollaender
Oak Ridge National Laboratory
Oak Ridge, Tennessee
and
University of Tennessee, Knoxville

Ronald F. Probstein
Massachusetts Institute of Technology
Cambridge, Massachusetts

E. S. Starkman
General Motors Technical Center
Warren, Michigan

Bruce L. Welch
Friends Medical Science Research Center, Inc.
and
The Johns Hopkins University School of Medicine
Baltimore, Maryland

Environmental Dynamics of Pesticides

Edited by

RIZWANUL HAQUE

Department of Agricultural Chemistry and
Environmental Health Sciences Center
Oregon State University
Corvallis, Oregon
currently
U.S. Environmental Protection Agency
Office of Pesticide Program
Washington, D.C.

and V. H. FREED

Department of Agricultural Chemistry and
Environmental Health Sciences Center
Oregon State University
Corvallis, Oregon

PLENUM PRESS · NEW YORK AND LONDON

Library of Congress Cataloging in Publication Data

Main entry under title:

Environmental dynamics of pesticides.

(Environmental science research; v. 6)
Proceedings of a symposium sponsored by the Division of Pesticide Chemistry, American Chemical Society, held during the 137th National American Chemical Society Meeting in Los Angeles, Apr. 1974.
Includes bibliographies, and index.
1. Pesticides—Environmental aspects—Congresses. I. Haque, Rizwanul, ed. II. Freed, Virgil Haven, 1919-   ed. III. American Chemical Society. Division of Pesticide Chemistry.
QH545.P4E48   574.5'222   74-28273
ISBN 0-306-36306-2

Proceedings of a symposium on Environmental Dynamics of Pesticides held during the 137th National American Chemical Society Meeting in Los Angeles, California, April, 1974

© 1975 Plenum Press, New York
A Division of Plenum Publishing Corporation
227 West 17th Street, New York, N. Y. 10011

United Kingdom edition published by Plenum Press, London
A Division of Plenum Publishing Company, Ltd.
4a Lower John Street, London W1R 3PD, England

Printed in the United States of America

Preface

Pesticides have played a significant role in increasing food production, and in view of growing worldwide food demand we can expect the use of these chemicals to increase. However, some of them have found their way into the biosphere and have been classified as persistent toxic chemicals. This has resulted in serious concern about environmental contamination. Since we are going to continue using chemicals, we should learn more about such aspects as their transport in the environment, the relationship of their physical-chemical properties to transport, their persistence in the biosphere, their partitioning in the biota, and toxicological and epidemiological forecasting based on physical-chemical properties. Environmental chemodynamics is the name given to a subject which deals with some of the above topics, utilizing the principles of such disciplines as chemistry, physics, systems analysis, modelling, engineering, and medical and biological sciences.

To ensure the safety of the environment, we must know more about the chemodynamical behavior of pesticides and related chemicals. The purpose of the symposium "Environmental Dynamics of Pesticides" was to explore the concept of chemodynamics as applied to pesticides and thus may help in developing the emerging field of environmental chemodynamics. The symposium was held during the 137th National American Chemical Society Meeting at Los Angeles, California, during April, 1974. The three sessions in the symposium were chaired by Drs. V.H. Freed, D.G. Crosby, and R. Haque.

This volume of the proceedings contains papers on behavior of chemicals in air, water, soil, and biota and includes such topics as modelling in the environment; photochemical behavior; adsorption, leaching, and breakdown; vapor loss; interaction with biological macromolecules; and detoxication by biota. The contributors are recognized experts in their fields and represent academia, industry, and government laboratories. We thank all the contributors for submitting their manuscripts. Without their interest and cooperation, it would have been impossible to complete this book. Each author should be credited for his own contribution.

We are grateful to the Division of Pesticide Chemistry of the American Chemical Society, its membership, and its officers for

sponsoring this symposium. The editing of the book has been supported partly by grants ES-00040 and ES-00210 from the National
Institute of Environmental Health.

        Thanks are due to Ms. Anna Moser and Ms. Connie Brown for
their excellent job of typing the entire book.

Corvallis, Oregon   R. Haque
                                     V.H. Freed

Contents

ENVIRONMENTAL DYNAMICS OF PESTICIDES: AN OVERVIEW

V.H. Freed and R. Haque, Department of Agricultural

Chemistry and Environmental Health Sciences Center

Oregon State University, Corvallis, Oregon 97331

Chemicals for many years have been an important means of controlling pests afflicting man, his animals, and food crops. Informed opinion is a consensus that with the growing world food crisis, chemicals for pest control will continue to be vital in production of food. This does not minimize other non-chemical approaches, or integrated pest control approaches, but rather recognize that pesticides remain in many instances our sole weapon of defense.

Beneficial though use of pesticides is, it is not without attendant problems. One of these problems is that not all of the chemicals will remain in the area of treatment. The physico-chemical properties of the substance, together with environmental transport processes result in a portion of the chemical released moving elsewhere in the environment. Indeed this has been one of the puzzling and even troublesome questions of our time. Traces of certain of the more persistent chemicals have been found well removed from areas of treatment, that is quite common to find water bodies near to, but not necessarily adjacent to treated areas contaminated with the material. Instances of long distance; i.e., several hundred miles, aerial transport have been reported.

The finding that pesticides were being transported from areas of treatment confronts us with a number of challenging scientific questions; among them:

1. What mechanisms are involved in this transport process, and how may they be accurately described and quantitated.

2.   What amount or proportion of the material that has been used is thus transported.

3.   What is the fate and behavior of the material during the transport process and upon arrival at the ultimate destination.

4.   What, if any, are the biological consequences.

5.   How to ensure that the analytical methods accurately identify and measure the low level residues.

Biologists and chemists for many years have studied fate and behavior of chemicals, but usually within a limited sphere and in only one component of the environment. More recently, scientists have essayed a broader more comprehensive study of these problems. These efforts have evolved into a systematic, interdisciplinary area of study that has come to be labeled chemodynamics. Concern in this symposia with environmental dynamics of pesticides is illustrative of the interest and approaches being used today. The field has three basic areas of concern; namely,

1.   An understanding of and model of the environment and the various interactions.

2.   A detailed knowledge of the chemical structure and properties of pesticides.

3.   Interaction of the chemical and its properties with elements of the environment as related to behavior, transport, and fate of the chemical.

Recognizing that there are four components of the environment; namely, atmosphere, lithosphere (soil), hydrosphere (water), and biosphere (living organisms), that specific parameters measure the properties of each component and that there is a dynamic equilibrium among them, permits a modeling of the transport processes and the distribution of the chemicals in the four components. Such distribution can be described with appropriate mathematical equations.

This symposium will concern itself with the behavior of pesticides in air, water, soil, and the biota. Rather specific attention will be given to the transport processes, physico-chemical properties of the compounds such as vapor pressures, solubilities, etc., that influence behavior and the mechanisms of partitioning of the pesticide into a particular component of the environment. Additionally, attention will be given to the fate of the chemicals as influenced by photochemical reaction, microbiological attack, and other types of reactions. What the symposium is attempting to do is rather new; namely, a systematic holistic approach to the dynamics of pesticides.

We believe this to be a step that will give us penetrating new insights into the problem, and lay a basis for further scientific development that will allow prediction of the behavior of not just pesticides, but any chemical that may get into the environment. Those making presentations are acknowledged experts in their field.

Elaboration of the knowledge of the dynamics of pesticides will lead not only to a better understanding of the fate and behavior of chemicals in the environment, but will point the development of improvement of our technology of the use of these compounds. This can mean safer, less environmentally contaminating practices, greater effectiveness, both of which should contribute to the production of food and the protection of man's health.

CHEMICALS IN THE AIR: THE ATMOSPHERIC SYSTEM AND DISPERSAL OF

CHEMICALS

R.L. Pitter and E.J. Baum

Oregon Graduate Center

Beaverton, Oregon 97005

INTRODUCTION

Field studies are required to validate any numerical model of pesticide dispersion. The various complications which may arise must be considered in the design of the validation study.

An analogous situation in the environment to pesticide dispersion is the dispersion of effluents from highway sources. A study of the nature of dispersion of these effluents can be used to better characterize the behavior of pesticide dispersion from crop spraying events. Several suggestions for control of pesticide dispersion are realized through application of the analogy.

DISPERSION MODELS

The local time rate of change of concentration of an atmospheric specie is equal to the sum of time rate of change of concentration of the specie due to advection, diffusion, chemical reactions, sources and sinks. Advection refers to transport of an ensemble of the specie due to the action of the wind. If all other terms are neglected, the advection term describes the motion of the ensemble (a puff or a plume) as it travels with the wind. The total time rate of change of concentration of a specie is the local time rate of change minus the advection term. The total time rate of change is mathematically expressed as:

$$\frac{dC}{dt} = \frac{\partial C}{\partial t} + \bar{u}\frac{\partial C}{\partial x} = \frac{\partial}{\partial x}[K_x\frac{\partial C}{\partial x}] + \frac{\partial}{\partial y}[K_y\frac{\partial C}{\partial y}] + \frac{\partial}{\partial z}[K_z\frac{\partial C}{\partial z}] + R + S, \quad (1)$$

where C is the specie concentration, $\bar{u}$ is the mean wind speed, K_x, K_y, and K_z are diffusion coefficients in the various coordinate directions, R is a generalized reaction term and S is a generalized source and sink term. The coordinate system is oriented with the x-axis in the direction of the mean wind and the z-axis vertical. The three diffusion terms relate the rate of concentration decrease due to volume expansion as the ensemble expands along each of the principal axis. Reactions, sources, and sinks are expressed only in general terms in equation (1). When the specific processes are identified, the proper relationships can be substituted.

The most widely used models for specie dispersion are Gaussian models. The Gaussian models are alike in the assumptions made re-garding dispersive mechanisms but differ in their source charac-terizations. The Gaussian puff model assumes a point source emitting a specie in one puff and follows the puff with time. The Gaussian line model assumes a time-independent emission from a line source. It incorporates the wind component normal to the source and assumes two-dimensional dispersion. Finally, the Gaussian plume model describes a point source emitting at a constant rate, Q, and as-sumes that locally steady state conditions exist. If a neutrally buoyant plume is emitted at ground level, and if the ground acts as a perfect reflector of the specie, equation (1) may be solved by Laplace transforms to yield:

$$C(x,y,z) = \frac{Q}{\pi\sigma_y\sigma_z\bar{u}} \exp\left\{-\left(\frac{y^2}{2\sigma_y^2} + \frac{z^2}{2\sigma_z^2}\right)\right\},$$ (2)

where

$$\sigma y = 2K_y x/\bar{u} \text{ and } \sigma_z = 2K_z x/\bar{u}.$$ (3)

Despite its simple derivation, the Gaussian plume model exhi-bits several interesting features of plume dispersion. The terms σ_y and σ_z are related to plume width and depth. From equations (2) and (3) it is seen that as diffusivity increases due to meteorologi-cal factors, the maximum specie concentration at some distance down-wind of the source is diminished, but plume width is increased. The exponential term in equation (2) describes binomial distribu-tions of specie concentration in the crosswind and vertical direc-tions.

Because of the simplifications involved, Gaussian models do not describe concentration dependence on topography, aerosol prop-erties, reactions, or weather conditions. The next section examines some of these complicating factors and discusses their effects on the model and on field study requirements for model validation.

COMPLICATING FACTORS

Turbulent Diffusion

The diffusion terms in equation (1) assume Fickian, or molecular, diffusion, whereby the mean square displacement of specie particles along any coordinate axis is a function of temperature, viscosity, and time. However, atmospheric diffusion of an ensemble of particles proceeds largely due to the action of turbulent eddies displacing portions of the ensemble. Figure 1 depicts the difference between molecular and turbulent diffusion acting on an ensemble. Turbulent eddies act in a discrete manner, displacing portions of the ensemble and allowing a much greater rate of diffusion. In order to justify the use of Fickian diffusion principles it is necessary to show that turbulent diffusion is isotropic, to determine the coefficient values from atmospheric observations, and to view concentrations predicted by equation (2) as time averaged values because of the discrete nature of eddies. Field experiments have verified that atmospheric turbulence is nearly isotropic in general, and empirical diffusion coefficients have been obtained for given localities. However, diffusion coefficients do vary with locality because of topographical features, and so field study is required prior to application of a model in all instances.

Since the diffusion parameter values are determined empirically, the model loses its theoretical significance. The only remaining

Molecular Diffusion   Turbulent Diffusion

Figure 1. Molecular and turbulent diffusion on an ensemble.

justification for using the model is its ability to match observed
results. As myopic as this approach may appear, it is fundamental
to the development of all atmospheric diffusion models to date and
is due to the complex nature of atmospheric turbulence.

In the process of validating a model, it is necessary to average the instantaneous values of the specie concentrations measured.
The averaging time should be equivalent to the transport time of
an air parcel traveling from the source to the receptor. Consequently, more distant receptors require longer averaging times. There
is a limit to the transport distance over which a model can be validated. This is because the mean wind speed and direction may change
during the sample collection period if the averaging time is very
long.

Topographical Features

The Gaussian models assume flat terrain. In actuality, this
assumption is rarely met. More typically, important topographical
features exist in the vicinity of the source in question. Buildings
and mountains serve to divert air flow, while mountain passes channel
air flow. Canyons, ravines, and valleys are subjected to a complex
pattern of air flow when prevailing winds approach them at oblique
angles. A helical spiral pattern is established with its axis
aligned along the longitudinal axis of the valley. The result of
air pattern complexity is to modify specie concentrations in certain
areas. Gaussian models are unable to correct for the effect of complex terrain, but some more advanced dispersion models approximately
treat them.

An important small-scale effect arises from the presence of
buildings on an otherwise flat terrain. The building disturbs the
flow locally and creates complex dispersion patterns in its vicinity.
The local specie concentrations are altered from expected values due
to the air flow pattern near the building. Although this effect is
small-scale, it will have important consequences if field sampling
sites are located in the vicinity of buildings. Such a validation
study would not necessarily yield the true specie concentrations.

Chemical Reactions

Gaussian models treat species as though they are inert gases.
Airborne pesticides are partitioned between the vapor state and the
adsorbed state on aerocolloids. As colloid material is added downwind of the source, transfer of pesticide from the vapor to the adsorbed state will occur and will serve to deplete the specie vapor
concentration. Chemical removal processes such as oxidation, hydrolysis, and catalytic decomposition may be promoted for the adsorbed

species. Photochemical processes may also be promoted. For instance, DDT undergoes photochemical conversion to form DDE in normal
sunlight when DDT is in the adsorbed state (Mosier and Guenzi, 1969).
This can be expected to occur on aerocolloids leading to the deple-
tion of DDT and an increase of DDE concentration downwind of the
source.

Aerosol Properties

For pesticide dispersion, adsorption of the specie on aerosols
may be pronounced, and consequently the aerosol properties must be
accounted for in a realistic model. Aerosols have finite settling
velocities conforming to Stokes' law. Aerosols smaller than about
two microns radius have very small settling velocities and may be
treated like gaseous particles with good accuracy. However, larger
aerosols have noticeable settling velocities and, as a consequence,
fall to the ground near the source.

There are numerous natural and anthropogenic sources of aerosols. Aerosols are continuously agglomerating and shattering, that
is, continuously evolving. Removal mechanisms such as fallout and
impaction are acting continuously. Also, during rain and snow, some
aerosols are removed by scavenging.

Some advanced dispersion models could be modified to account
for the different fall velocities of various size aerosols, but the
treatment of the evolution of aerosols has yet to be incorporated
in any models. Aerosol evaluation may be neglected if the rate of
evolution is slow compared to the specie transport time.

Meteorological Factors

Most Gaussian models assume that the wind speed and direction
are everywhere constant, independent of time or position. However,
wind speed and direction typically vary with height in the lowest
few hundred meters of the atmosphere. This phenomenon, the Ekman
spiral, is an effect of friction at low levels of the atmosphere.

The effect of the Ekman spiral is to cause higher dispersing
ensembles to advect in different directions at different speeds,
thus somewhat enhancing the rate of dispersion of the specie. Because of the complex nature of dispersion under these conditions,
validation studies must be accompanied by crosswind analyses to
determine where the plume line is located.

Diurnal wind changes are also typical. Sea breezes are common
near shores of lakes and oceans, and have pronounced diurnal patterns. Narrow valleys also experience diurnal ebb and flow of air

motion. In general, air is quite stable and calm in the early
morning, but becomes more turbulent and gusty in the afternoon.
These effects greatly alter the eddy diffusivity coefficients. Gaussian models employ fixed eddy diffusivity coefficients and cannot
be accurately applied in such situations.

In rainy weather several mechanisms act to remove gases and
aerosols from the atmosphere. Some aerosols act as condensation or
ice formation nuclei to form cloud particles. These aerosols are
removed from the atmosphere when the water drops or ice crystals
precipitate.

Scavenging also occurs. The precipitating droplets could col-
lect aerosols and trace gases and quickly transport them to the
ground. It is commonly observed that the air is cleaner after rain
than prior to it, and a portion of the cleansing is due to scavenging.
However, a greater cleansing role is played by the ventilation which
occurs in rainy weather. Vertical air motion, or convection, acts
to disperse the specie to greater altitude, thereby significantly
diluting it, but also simultaneously allowing its transport over
great distances. Sensitive measurements have detected the presence
of pesticides in Antarctica. It is probable that the pesticides
were transported there at high levels in the atmosphere and brought
to the surface through diffusion or convection.

OTHER DISPERSION MODELS

Those effects resulting from topography, chemical reactivity,
aerosol dynamics, and meteorological factors cannot be accounted
for by simple Gaussian models. More elaborate Gaussian models can
account for parameterized wind shear. Although we have emphasized
Gaussian models to promote transparent concepts, attention should
be given to advanced numerical modeling techniques. These have developed considerably toward a more realistic treatment of the problem.
Although many such models have been devised, there are only three
major classifications.

Mass Conservation models employ differential equations relating
to the dispersion of a specie in finite-difference form, and formu-
late the problem in a manner so as to conserve the total mass of
the specie. The atmosphere is divided into a system of cells, and
the model follows the specie as it advects and diffuses from one
cell to the next.

Particle-in-Cell models obtain a more realistic description of
specie diffusion than do mass conservation models. Essentially, the
program numerically generates a statistically significant ensemble
of tracer particles of the specie which are emitted from the source
and are followed as they disperse according to randomly generated
turbulent eddies.

Rather than partition air space into cells which are fixed
with respect to topographical features, certain simplifications
can be realized through the use of a Lagrangian system, whereby
the cells are allowed to move with the air. Diffusion over a small
time interval is not treated by differentiation but by integration,
which is inherently more stable on the computer.

At the risk of being redundant, it is emphasized that any model,
regardless of how advanced, must be validated. The preceding dis-
dussion of complications which may arise during validation studies
is applicable regardless of the model being tested.

HIGHWAY ANALOGY

Besides briefly presenting dispersion models and complicating
factors, the authors wish to show how some of their field studies
of dispersion of highway-related emissions can be applied to pesticide dispersion to yield significant results.

Like crop spraying events, automobiles emit a ground level,
neutrally buoyant plume composed of gases and aerosols. Traffic-
generated turbulence entrains road dust into the air, and there is
a background aerosol loading upon which the traffic emissions are
superimposed.

Figure 2 was compiled on the basis of field sampling in the
vicinity of highways to illustrate a simple qualitative model of how
air quality is affected in surrounding areas.

Closest to the highway, the aerosol loading is greatest, but
it falls off rapidly within the first few minutes transport time.
This indicates that the majority of aerosols in this region are
greater than a few microns in size and fall out very rapidly.

Further from the highway, impact on visibility reduction is
small and the aerosol mass loading due to the highway is small.
This level does not greatly change with increasing transport time
to an hour or more from the source because the aerosols are predominantly submicron in size and possess negligible fall velocities.

A secondary maximum in aerosol loading has been noted by means
of nephelometry (visibility reduction) measurements downwind of
communities. This maximum is due to the secondary formation of
aerosols from emitted trace gases. The position of the peak is
dependent on windspeed and reaction rate, and thus on available
radiation. Peak height is dependent on source strength and reaction efficiency.

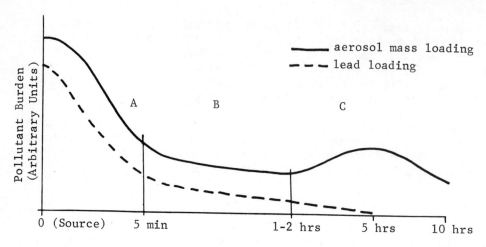

Fig. 2.   Pollutant Transport Time

Neglecting for the moment the obvious discrepancy in source geometry between a highway and crop spraying, it is interesting that lead should behave like many pesticides in the analogy. Both species are relatively inert, both are emitted as gases and aerosols and both may be strongly adsorbed on aerosols from the vapor phase.

Our studies of aerosol lead concentrations downwind of highways show a similar distribution to that of total aerosol concentration in the first two zones of the model (Figure 2). Since no lead aerosol is formed by secondary reactions with trace gases, the secondary peak is not present. The highest concentrations of lead aerosols are found only in the immediate vicinity of highways. The rapid decrease in lead concentration with transport time from the source indicates much deposition. On the basis of chemical analysis, we find that up to 50% of airborne lead at highways is carried by aerosols greater than a few microns in size. The remaining particulate lead is in submicron aerosols and consequently remains airborne for much longer times, dispersing to great distances.

Chemical element analysis of aerosols reveals that there are several paths of lead through the environment. If automotive lead, which is emitted predominantly in two size ranges (one submicron and one supermicron) is the only source of lead, then there would be an expected close correlation between lead and bromine, which is also predominantly emitted by automobiles. However, our results show that even in areas where the automobile is the only significant lead source, lead is not well correlated with bromine alone in the supermicron range, but rather a multiple linear correlation of lead with bromine and silicon, a typical abundant soil element, yields an extremely good correlation coefficient, indicating that super-micron lead containing aerosols originate as automobile emissions and soil dusts. Our analyses indicate about a 50/50 division of contribution from the two sources. Consequently, lead accumulates in road dust during dry periods of weather, and traffic-generated turbulence serves to entrain the dust in air and to disperse it. A similar mechanism occurs in the submicron size range, but to a lesser degree because of the lower deposition rate of smaller aerosols near the roadway.

Our highway studies can be used as an analogy to crop spraying events. Pesticides are sprayed as liquid or solid aerosols, and contain a certain amount of pesticide in the volatilized gaseous phase. The aerosol pesticides are predominantly large--up to 5000 microns--although natural processes create smaller particles as well. When dispersed in the atmosphere, the large particles will quickly settle to the ground, while particles less than a micron in size will disperse in the wind. If a liquid aerosol is used, com-plications may arise through evaporative reduction of the aerosol size. The gaseous pesticide portion will seek equilibrium with the aerosol portion, and the presence of natural aerosols enhances the conversion of pesticide from the gaseous to the adsorbed state. Finally, turbulence generated by the spraying process will entrain soil dusts into the air, creating a greater aerosol mass loading.

Beyond the source boundaries one can expect predominant pesti-cide transport associated with that of submicron aerosols. If a single field is sprayed, there will be a faster decrease in airborne concentrations away from the source than in the highway study due to the dissimilar source geometries involved. However, if widespread spraying is conducted, there may be a generally higher airborne pes-ticide concentration due to the multiplicity of sources.

CONCLUSION

Dispersion models depict the action of advection, diffusion, sources, sinks, and in some cases, reactions. Because of the nature of turbulent diffusion in the atmosphere, models can only be expec-ted to yield average concentrations of species, and must be validated by field studies. Field studies for model validation must be well planned to obtain useful data. Meteorological and topographical complicating factors must be considered.

Highway studies offer an analogy for dispersion of pesticides in the atmosphere from crop spraying events. Both cases consist of

ground level, neutrally buoyant plumes, and aerosols disperse similarly under such conditions. The current highway study has noted
that much lead is concentrated near the source during dry periods,
and is then entrained with soil dust for dispersion. Such may also
be the case with sprayed pesticides.

Several suggestions for reducing pesticide dispersion include:
1) use of low volatility pesticides; 2) use of large spray aerosol
sizes; and 3) prior wetting of treated areas to suppress turbulent
entrainment of soil dust.

SUMMARY

During crop spraying events, pesticides are introduced into
the atmosphere as solid or liquid airborne matter, termed aerosols,
and as volatilized gases. The detailed nature of pesticide dispersion in and removal from the atmosphere is governed by many factors.
Atmospheric turbulent diffusion is a complex process which determines the rate of dispersion of the plume. In order to reasonably
model plume dispersion, it is necessary to empirically correlate
the model's diffusion parameters with results of atmospheric observations of plume dispersion. Topographical features often act
to enhance or reduce concentrations from those predicted by simple
calculations. The detailed exchange of pesticides between the
aerosol and vapor phase depends on properties of the pesticide and
surface properties of the aerosols. Other aerosol properties
govern the transport of adsorbed pesticides. Meteorological parameters also considerably influence pesticide transport.

ACKNOWLEDGMENT

This study was supported in part by the Highway Division of
the State of Oregon and by the Federal Highway Administration under
Air Quality Study 5149-621-10. The information contained herein
is the responsibility of the authors.

REFERENCES

Dabberdt, W.F., F.L. Ludwig, and W.B. Johnson. 1972. Urban dif-
    fusion of carbon monoxide. H.W. Church et al., editors.
    NTIS Pub. CONF-711210, 145-151.

Lamb, D.V., F.I. Badgley, and A.T. Rossano, Jr. 1973. A critical
    review of mathematical diffusion modeling techniques for pre-
    dicting air quality with relation to motor vehicle transpor-
    tation. NTIC Pub. PB-244 656.

Mosier, A.R., W.D. Guenzi, and L.L. Miller. 1969. Photochemical
    decomposition of DDT by a free radical mechanism. Science 164:
    1083.

Slade, D.H., editor. 1968. Meterology and Atomic Energy: 1968.
    Atomic Energy Commission, Washington, D.C.

Sutton, O.G. 1953. Micrometeorology. McGraw-Hill, New York.

DETERMINATION OF PESTICIDES AND THEIR TRANSFORMATION PRODUCTS IN AIR

James N. Seiber and James E. Woodrow
Department of Environmental Toxicology
University of California, Davis, California 95616

Talaat M. Shafik
Pesticides and Toxic Substances Effects Laboratory
U.S. Environmental Protection Agency
Research Triangle Part, North Carolina 27711

Henry F. Enos
Division of Equipment and Techniques
U.S. Environmental Protection Agency
Washington, D.C. 20460

The atmosphere has been postulated as a major route for the widespread distribution of the more persistent pesticides in the environment (Risebrough et al., 1968; Woodwell et al., 1971). This has been supported by a growing reservoir of data reporting the presence of residues in air samples from areas far removed from normal sites of application. Risebrough et al. (1968), for example, found an average concentration of DDT-related pesticides and dieldrin of 7.8×10^{-14} g/m^3 in dust collected at Barbados in 1965-1966. A later study (Seba and Prospero, 1971) revealed average concentrations of aerosol-borne DDT and DDE approximately three times higher in tradewinds sampled at Barbados in 1968. In a more recent study (Bidleman and Olney, 1974) average DDT levels of approximately 20×10^{-12} g/m^3, 100 times those of the earlier studies, were reported from air sampled at and near Bermuda.

These figures, while developed with a variable air mass, show that our knowledge of concentrations in the air of even the most uniformly distributed pesticides is far from perfect. This is further underscored in the data from air samples collected over land (Abbott et al., 1966; Tabor, 1965; Jegier, 1969; Stanley et al., 1971; Södergren, 1972; Yobs et al., 1972; Zweig et al., 1972).

Despite the quantitative variability, the fact of the existence of
residues in the air may be considered proved. This fact should
stimulate development of more reliable quantitative methods for
routine use, so that the extent of the contamination, and thus its
importance, can be thoroughly examined.

ORIGIN OF PESTICIDE RESIDUES IN AIR

Simple logic suggests several potential sources of airborne
residues (Figure 1). Drift during normal application may lead to
movement of chemicals, often several miles from the source (Akesson
and Yates, 1964; Yule and Cole, 1969; Yule et al., 1971). The
relative importance of drift to total atmospheric pesticide concentrations is probably not great, however, since it involves
primarily particulate material able to settle out in a relatively
short time. Evaporation and erosion losses from application sites
and nearby areas have been suggested to be the more important
sources (Freed et al., 1972). Estimates of evaporative losses of
a few percent (Caro et al., 1971) to as much as half (Hartley, 1969;
Lloyd-Jones, 1971) of agriculturally-applied chlorinated insecticides have been made. The importance of vapor pressure, soil type,
and moisture on this important physical process has been discussed
in detail elsewhere (Spencer, 1970; Spencer and Cliath, 1974).
The contribution of erosion, primarily involving dust and soil particles, is even more difficult to estimate; the sparse available
data is quite suggestive, however, and indicates a need for further
study of this factor (Cohen and Pinkerton, 1966; Freed et al., 1972).

Other sources of localized importance may derive from manufacturing and formulation sites, and industrial and agricultural
wastes. Furthermore, revolatilization of precipitated or settled
residues from surfaces far removed from the original source undoubtedly serves as a route of re-introduction; a "dislodgeable"
concept similar to that of Gunther et al. (1973) may be useful to
differentiate these residues from soil and tissue incorporated or
bound chemicals.

FORM AND FATE

The form of residues in the atmosphere is of importance from
both the analytical and toxicological viewpoint. Initially present
as both vapors and associated with particulate matter, residues
should distribute themselves between these phases in accordance
with their own adsorptive and volatility characteristics, and the
properties of the atmosphere in which they reside. The available
evidence indicates that both forms contribute substantially to
the residues observed over land, though the proportions vary greatly

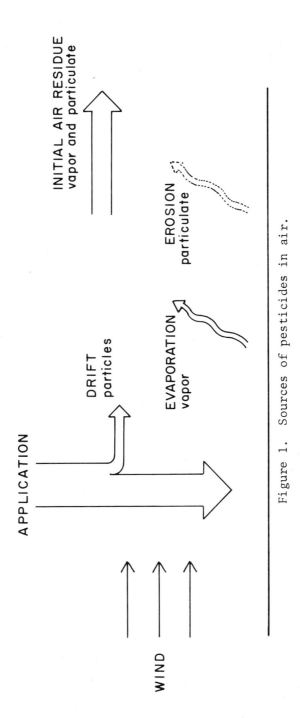

Figure 1. Sources of pesticides in air.

with the sampling site (Atkins and Eggleton, 1970; Stanley et al., 1971). On the other hand, chlorinated compounds determined in sea air near Bermuda were almost exclusively in the vapor form, though the methodology employed leaves room for some doubt in this conclusion (Biddleman and Olney, 1974). In short, this important question has not been satisfactorily answered to date. Its resolution represents a worthwhile goal of future air monitoring studies.

Once present in air, residues are subject to transformation and removal, primarily by precipitation. Chemical transformation, in both vapor and particulate forms, appears as a major dissipation route, given the excess of oxidant and sunlight, and the existence of catalytic surfaces and other reactants available to both forms (Crosby et al., 1974a). The possibility of chemical reactions in air multiplies the number of pesticide-derived products for determination as well as the potential non-pesticide interferences.

IMPORTANCE

The Federal Environmental Pesticide Control Act (1972) explicitly charges that monitoring of pesticides in air, as in other components and compartments of the biosphere, should be undertaken. Several reasons exist for implementing this activity:

1. The air represents an as yet ill-defined pesticide exposure route. For man, levels found to date in ambient air samples have been far below the threshold limit values of airborne contaminants adopted by the American Conference of Governmental Industrial Hygienists (Yobs et al., 1972), and are minor relative to the intake from total diet studies carried out by Duggan and Lipscomb (1969; Stanley et al., 1971). There are, however, dangers involved in extrapolating intake data to a population already exposed to a variety of pollutants. Furthermore, since rather large variations in concentration have been observed in levels measured to date, the possibility of significant exposure for many individuals, particularly farmworkers and rural dwellers, cannot readily be dismissed (Jegier, 1969). For wildlife and lower organisms, the exposure picture is even less clear.

2. The air may, as many have suggested, represent a significant route of pesticide transport and distribution. Systematic analyses of atmospheric samples from many locations may serve to correlate areas of higher and lower concentration levels than the average with meteorological data, such that patterns of contamination may be recognized.

3. Given accurate data, air monitoring could aid in assessing the results of changing pesticide use patterns on environmental

residues. The recent trend in usage from more to less persistent chemicals should eventually be reflected in lower average air levels (Woodwell et al., 1971).

4. At present, little is known of the nature of pesticide transformation products in air. In at least one case, a principal transformation product (DDE) may enjoy a significantly greater atmospheric mobility than its parent (DDT) (Spencer and Cliath, 1972). That this situation may exist with other pesticide trans-formation products--formed in condensed phases or directly in air--suggests the prudence of atmospheric surveillance.

5. From an academic standpoint, defining the role of the atmosphere in the fate of pesticides should lead us closer to the eventual goal of accountability for all potentially deleterious chemicals released intentionally to the environment.

ANALYTICAL METHODOLOGY

The determination of pesticides in air proceeds, as in most analyses, through discrete steps of sampling, extraction, cleanup and fractionation, resolution, and determination. Sampling in air studies refers to selecting and gathering representative particulate matter for analysis of suspended pesticides, or a representative volume of air for pesticide vapor analysis. The former usually in-volves some form of filtration, retained particles constituting the sample for analysis. Loss from the filtered particles, primarily by volatilization to the airstream, seems likely to occur, the extent being governed by the pesticides polarity and volatility, the adsorp-tive nature of the surface, and the air flow rate and temperature. Furthermore, loss of small particles through the filter is bound to occur. Use of glass cloth and glass fiber mat filters capable of efficiently trapping particles to at least 0.1 micron in diameter serves to minimize this.

With vapors some form of concentration, often by liquid ex-traction or adsorption, is employed simultaneously with the sampling step. This sampling-extraction is often referred to simply as "sampling," and it is characterized by the term "sampling efficiency." In fact, "sampling efficiency" reflects not only those losses in-curred in sampling, through incomplete extraction, but also subse-quent loss from the sampling medium by revolatilization or chemical degradation. Many of the difficulties encountered in determining sampling efficiency are thus analogous to those in determining extraction efficiencies of residues from condensed substrates (van Middelem, 1971). Most efforts for efficiency determination to date have involved spiking air with known amounts of pesticides as dusts or vapors and determination of overall recovery. The ability to approximate conditions of form and concentration which exist in actual samples by spiking are somewhat questionable.

"Sampling" is obviously a key step in determining pesticides in air. The efficiency of sampling will reflect the physical and chemical properties of both the pesticides trapped and the sampling medium. To summarize, an ideal sampling medium should:

1. Efficiently trap pesticides of interest, from a sufficient volume of air sampled at a suitable rate, with a minimum of analytical interferences trapped simultaneously;

2. Retain trapped pesticides for the duration of sampling;

3. Be inert itself toward sampled pesticides, and minimize their reaction with oxygen, water, and other potential reactants present in air during sampling;

4. Readily release pesticides in a form amenable to analysis on post-sampling workup;

5. Contribute no interference itself to subsequent analytical steps, either as contaminants present originally in the medium or formed by its degradation during the sampling process;

6. Be inexpensive, available, easily handled, etc.

The remainder of the analysis, extraction of the pesticides from the sampling medium, their cleanup, resolution, detection, and measurement, is analogous to more traditional residue methodology. Multiresidue methods capable of processing all, or at least most, of the pesticide types and their derived products of interest are essentially those used in determining residues from condensed phases. Since the bulk of organic matter extracted from air is relatively small compared with that extracted from living tissue and soil, micromethods of cleanup and fractionation, such as employed in the scheme described by Sherma and Shafik (1974), are particularly suitable for air samples. The limit of detectability attainable for any given pesticide in air will reflect the efficiency (and background) in both the sampling and subsequent analytical steps.

Liquid Absorption Methods

A number of methods have been employed for trapping pesticides from air (see summary in Miles et al., 1970); for the most part, these methods were applied to only one or a few chemicals, often at or near application sites where relatively high concentrations

exist. In many earlier studies, attention was largely confined to
residues of the persistent chlorinated insecticides rather than a
cross-section of the different types of pesticides in common use
(Tabor, 1965; Abbott et al., 1966).

One of the first attempts at method development aimed at a
variety of compounds in ambient concentrations was that of Miles
et al. (1970). Ethylene glycol was employed as solvent in Greenburg-Smith impingers through which air was sampled at 10.3 lpm.
This device was tested by connecting two impingers in series and
found to work well for both vapors and dusts (Table 1). Losses
increased on prolonged sampling; they were greater with more volatile compounds such as diazinon. The limits of detectability for
several chlorinated and organophosphate insecticides were 1-10 ng/m^3,
based on GLC response to standards. No cleanup was involved since
spiking levels were high.

Stanley et al. (1971) employed a larger version of the
Greenburg-Smith impinger, charged with 100 ml of hexylene glycol
rather than 25 ml of ethylene glycol (Figure 2). An alumina adsorption tube was added, to trap pesticides not retained in the
impinger. The flow rate was 29 lpm, providing an air sample of
$\sim$40 m^3 in 24 hours. Spiking the inlet, without the glass cloth
filter in place, gave good recoveries of several pesticides in the
hexylene glycol, after correction for unvaporized pesticides in the
inlet. The tests did, however, show the need for the alumina backup. A Florisil cleanup-fractionation was included; poor recoveries
for several compounds through the analytical system necessitated
the use of rather large correction factors (up to 2.8, for malathion).

The method was applied to ambient samples collected at several
locations. A partial list of the resulting data (Table 2) showed
significant amounts of trapped chemicals present in all three sampler
components. For one unusual sample, the filter (f) contained nearly
ten times as much p,p'-DDT as did the impinger (h) and adsorbent (a).
The limits of detectability were stated as 0.1 ng/m^3, but no data
were presented to support this.

The methods of Miles et al. (1970) and Stanley et al. (1971)
became, with some modification, the basis of a two-year monitoring
program carried out by the Division of Community Studies of EPA
(Enos et al., 1972). Ethylene glycol was chosen as solvent, giving
less interfering background than hexylene glycol. The alumina
backup was eliminated, further reducing interferences. Use of
Mills-Onley-Gaither FDA cleanup led to satisfactory recoveries,
eliminating the need for correction factors. The limits of detectability for quantitation were approximately 1-4 ng/m^3 for most
chlorinated and organophosphate pesticides. Levels as low as 0.1
ng/m^3 were detectable for many compounds.

TABLE 1

Efficiency of Greenburg-Smith Impingers for Trapping Parathion
and Diazinon in 25 ml Ethylene Glycol (10.3 1pm)[a]

			Percent Trapped		
Compound	Form	Period of Run (Min)	Impinger I	Impinger II	Cold Trap
Parathion	Vapor	300	90.2	9.8	0
	Dust	4	99.9	0.1	0
Diazinon	Vapor	60	86.0	12.3	1.7
	Vapor	240	79.5	18.2	2.3
	Dust	4	99.5	0.5	0

[a] Data selected from Miles et al. (1970).

A number of details remained to be worked out with the impinger
system. Ethylene glycol in the larger impinger, charged with 100
ml of solvent, had never been rigorously tested in the laboratory
with the large number of chlorinated, organophosphate, and carbamate pesticides of monitoring interest. Furthermore, no spiking
studies had ever been carried out with pesticide levels near those
present in ambient air. Thus the ethylene glycol system was subject
to even more scrutiny. The testing protocol of Miles et al. (1970),
two impingers in series with spiking through use of a heated U-tube
located just prior to the first, was followed. A silica gel cleanup
and fractionation (Sherma and Shafik, 1974) was substituted for
Florisil, to allow for determination of carbamate insecticides.
The sampling rate (23 1pm) and duration (12 hr) were essentially
those used in the monitoring program.

The results (Table 3) showed method recoveries of >66% for all,
and trapping efficiencies >48% in the first impinger for all but
aldrin of 23 pesticides tested. Spiking levels ranged from 0.5 to
23 ng/m^3. From the quantities found in the second impinger, recoveries were seen to be generally lowest with the least polar and
most volatile chlorinated insecticides α-BHC, aldrin, and p,p'-DDE.
Limits of detectability for quantitation were calculated based on
GLC response of standards above the reagent background (Table 4).
They ranged from 0.1 ng/m^2 for α-BHC and lindane to 16 for carbaryl.
These limits are not corrected for impinger or method losses which

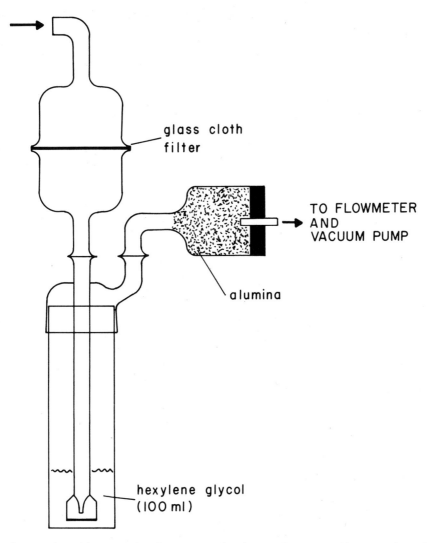

Figure 2. Filter, impinger, and adsorption sampling train of
Stanley et al. (1971).

TABLE 2

Distribution of Pesticides in Collection Media[a]

Locality	Sample	Pesticide Level, ng/m^3		
		p,p'-DDT	o,p'-DDT	p,p'-DDE
Baltimore	109f	0.9	0.5	--
	109h	1.0	0.5	2.1
	109a	0.9	--[b]	--
Iowa City	90f	1.7	0.9	1.7
	90h	0.8	0.5	--
	90a	1.2	0.3	--
Salt Lake City	123f	1.8	--	--
	123h	3.9	--	--
	123a	0.8	0.1	--
Stoneville	113f	417.0	80.0	7.1
	113h	52.0	20.0	2.4
	113a	19.5	8.0	--

f = filter   h = hexylene glycol   a = alumina

[a] Data selected from Stanley et al. (1971).
[b] None detected.

would serve to raise most values by approximately 50-100%. Furthermore, the background from actual air samples can easily raise limits several-fold, particularly for those chlorinated insecticides of short retention times detected by electron-capture. Limits for carbamates were much higher than those for the chlorinated and organophosphate pesticides from the reagent background developed during derivatization with pentafluoropriopionic anhydride (Shafik and Mongan, 1972).

TABLE 3

Recovery of Pesticide Vapors in 100 ml Ethylene Glycol
Impingers (16.5 m³ Air Samples, 23 lpm)

	Added to U-Tube	% Trapped In Impinger		Recovery (%) Thru Method Only
		I	II	
α-BHC	8 ng	64	30	73
Aldrin	10	20	40	86
p,p'-DDE	20	80	29	94
o,p'-DDT	60	77	21	98
p,p'-DDD	80	72	16	95
p,p'-DDT	80	78	18	97
Lindane	8	71	16	66
β-BHC	80	56	12	75
Hept. Epox.	18	66	20	84
Dieldrin	40	75	18	86
Endrin	80	67	14	89
Ronnel	84	62	26	85
M. Parathion	162	80	16	88
E. Parathion	208	90	18	90
Trithion	376	48	10	78
Ethion	246	70	16	93
Diazinon	54	68	18	80
Malathion	230	60	14	102
Baygon	200	51	19	96
2,3,5-Landrin	200	52	22	99
Carbofuran	400	52	16	101
Carbaryl	400	76	16	110
Mesurol	400	61	14	101

The retention of pesticides was tested by spiking the ethylene
glycol in the first impinger directly with standards, then drawing
air through for 12 hours at 23 lpm. Recoveries from the first
impinger in this case were slightly higher for all compounds than
when the pesticides were introduced as vapors; thus revolatilization is not likely to lead to significant losses (except with α-BHC
and aldrin) and little hydrolytic or oxidative losses occur in the
impinger solution.

TABLE 4

Estimated Detectability Limits (ng/m^3) for Ethylene
Glycol Impinger (16.5 m^3 Air Sample)[a,b]

α-BHC	0.1	Ronnel	0.4
Aldrin	0.2	M. Parathion	0.7
p,p'-DDE	0.3	E. Parathion	0.9
o,p'-DDT	0.6	Trithion	2
p,p'-DDD	0.8	Ethion	1
p,p'-DDT	0.9	Diazinon	0.3
Lindane	0.1	Malathion	1
β-BHC	1	Baygon	3
Hept. Epoxide	0.3	2,3,5-Landrin	4
Dieldrin	0.4	Carbofuran	10
Endrin	0.7	Carbaryl	16
		Mesurol	4

[a] Based on GLC peaks giving 10% full scale deflection at attenuation 320 (electron-capture) or a signal to noise ratio of 4 at
attenuation 32 x 10^3 (flame photometric) above the background
of reagent blanks.

[b] No correction for trapping or method efficiency.

All in all, the ethylene impinger system traps pesticide
vapors and, from the data of Miles et al. (1970), particulate-
bound pesticides with high efficiency. It does, however, have its
limitations: the glassware is expensive and fragile, and flows are
limited to around 30 lpm. In the EPA monitoring system the ethylene glycol from four impingers operated for 12 hours each was
pooled to provide a sufficient sample (80 m^3 for 24 hours) for
sensitivity and confirmation. The use of a solid sampling bed
appears an attractive alternate to the impinger system by allowing
for increased flow rates, shorter sampling periods, and simpler
collector design.

Solid Media as Trapping Agents

Several types of solids may conceivably be used for sampling
purposes. Table 5 summarizes some of those which have been reported
for sampling pesticides in air. The following discussion will
center on those which have been examined in sufficient detail for
evaluation of their potential for ambient monitoring.

TABLE 5

Some Reported Solid Sampling Media

Inorganic Adsorbents

    Florisil   Yule et al., 1971
    Alumina   Stanley et al., 1971

Organic Polymeric Adsorbents

    Chromosorb 102   Thomas and Seiber, 1974
    Polyurethane   Bidleman and Olney, 1974

Liquid Phases on Solid Supports

    Cottonseed Oil on Glass Beads   Bjorkland et al., 1970
    Glycerine on Nylon Nets   Risebrough et al., 1968
    Ethylene Glycol on Nylon Nets   Tessari and Spencer, 1971
    SE-30 on Nylon Nets   Södergren, 1972
    Polyethylene on Silica Gel   Herzel and Lahman, 1973

Bonded Liquid Phases

    Octadecylsilicone on Chromosorb   Aue and Teli, 1971

One of the first serious attempts to develop a high volume sampling method for a variety of pesticides employed 3 mm glass beads coated with 3 ml cottonseed oil for collection (Bjorkland, Compton, and Zweig, 1970; Compton and Bjorkland, 1972). The beads were held in a glass adapter through which air was pulled at 10 cfm, approximately ten times the rate of the large Greenburg-Smith impinger. A single 24 hour collection thus afforded a 400 m^3 sample. Following collection, the beads were Soxhlet-extracted with hexane, the pesticides partitioned to acetonitrile, and determined without further cleanup by GLC.

When DDT-containing aerosols were introduced directly to the coated beads, high collection efficiencies resulted. For vapors, the collection efficiency exceeded 70% for eight compounds during two hour sampling tests. The limits of detectability were estimated as ranging from 0.18 (heptachlor) to 1.7 ng/m^3 (2,4-D), from the GLC responses of standards. While this system did show considerable promise in the limited testing carried out, it was not without drawbacks. Prolonged sampling of air results in partial breakdown of the cottonseed oil, with the formation of electron-capturing interferences and the occurrence of emulsions in extracts (R.G. Lewis, personal communication). Furthermore, the hexane-acetonitrile partition is not entirely efficient in separating cottonseed oil from the pesticides of interest; the oil which remains in the acetonitrile phase may interfere in subsequent cleanup and GLC analysis.

A similar approach has been employed by others for static sampling (Risebrough et al., 1968; Tessari and Spencer, 1971; Södergren, 1972). That of Tessari and Spencer (1971) is typical; nylon chiffon was saturated with 10% ethylene glycol in acetone, then suspended in a wooden frame to provide 0.5 m^2 total area. The resulting air sampling device is noise-free and requires no electrical hookups. Side by side comparison with two ethylene glycol impingers showed, after four hours exposure (2.5 m^3 of air processed by the impingers), about four times as much DDT in the screen as found in the impinger solution (Table 6). This method appears quite suitable for comparison of exposures from one enclosed location to another; it does not, however, allow calculation of concentrations. When employed out of doors, variations in wind direction and speed may make comparisons more difficult.

A drawback of all liquid coating methods lies in the extraction of relatively large amounts of oil along with the pesticides of interest. Aue and Teli (1971) proposed the use of liquid phases bonded to the support to overcome this problem. Their method, which utilized octadecylsilicones bonded to Chromosorbs, was found to be quite efficient in trapping a number of common air pollutants, including a few chlorinated insecticides.

A number of solids and coated solids were examined for pesticide sampling (Table 7), in the hopes that one or a few would emerge

TABLE 6

Comparison of 0.5 m^2 Nylon Chiffon
Screen and Two Greenburg-Smith Impingers[a]

	DDT Collected	Total μg
Screen	18.992 μg/m^2	9.461
Impingers	1.006 μg/m^3	2.487

[a] Tessari and Spencer (1971).

TABLE 7

Some Solid Trapping Media Screened, Arranged in Approximate
Order of Suitability for Pesticide Sampling

Adsorbents	Coated Chromosorb A
Silica gel	5% Paraffin oil
Chromosorb 102	20% QF-1
Florisil	20% Paraffin oil
Chromosorb A	1% Paraffin oil
Chromosorb W acid-washed	5% DC 200
Chromosorb W acid-washed, DMCS-treated	20% Carbowax 6000

equal or superior to the ethylene glycol impingers in efficiency
and background. In preliminary tests at low flow rates silica gel,
Florisil and Chromosorb 102 (cross-linked polystyrene) were the
leading candidates of those solids tested. Chromosorb A had some
retentive properties in its own right and several mechanical advan-
tages (large particle size, high capacity for liquids) which led to
its use as support for testing a number of liquid phases. A 5%
coating of paraffin oil performed best in both sampling efficiency
and electron-capture background. Several other liquids showed some
promise but no advantage over paraffin oil.

Comparison of 20/30 mesh Chromosorb A with an equal volume of
3 mm glass beads clearly showed the superiority in trapping efficiency of the former as a support for paraffin oil. When similar
tests were carried out with cottonseed oil and ethylene glycol,
efficiencies with Chromosorb A always exceeded those with the glass
bead support. Among the three liquid phases, paraffin oil and
cottonseed oil had equivalent trapping properties; the former,
however, had the lowest electron-capture background.

For further testing, a 5 cm diameter filter cup, the same as
used in the EPA monitoring program, was fitted with a wire screen
and filled about halfway with the solids to be tested (30 g 20/40
silica gel deactivated with 20% water, 30 g 5% paraffin oil on
20/30 Chromosorb A, or 5 g 60/80 Chromosorb 102). The flow restric-
tion at 23.5 lpm was no greater through this setup than through the

large ethylene glycol impinger. An impinger backup was used, to
determine amounts of pesticides not trapped by the solid. The
setup was that of Figure 2, without the alumina adsorption tube.
Samples were solvent-extracted, and cleaned up and fractionated on
silica gel as in the impinger tests described earlier. The efficiencies observed with silica gel and 5% paraffin oil on Chromosorb
A toward most of the 23 pesticides were quite high (>47%) when
tested under the same protocol as described for ethylene glycol
(Table 3). Exceptions were aldrin, essentially zero trapped in
both filters, and Trithion and mesurol (zero trapped on paraffin
oil, ∿20% on silica gel). The loss of aldrin was not too surprising
since this compound had been trapped poorly by the impinger method.
Since Trithion and mesurol both contain oxidizable sulfur, we tend
to ascribe the low efficiencies observed with these two compounds
to oxidative losses on the filtering media surface.

Reagent background from silica gel was slightly higher than
that from ethylene glycol, even though the former was purified by
Soxhlet extraction and conditioning at 175° overnight. A more
serious problem arose in the larger amount of electron-capturing
background trapped from the air by silica gel. This most likely
reflects the greater trapping efficiency of this solid for air
constituents not trapped well by ethylene glycol. While high
trapping efficiency seems to be a desirable goal in method design
generally, it appears as a distinct disadvantage for determinations
of those chlorinated pesticides having short GLC retention times.
Use of a Cl-selective detector could, perhaps, alleviate this
problem.

Chromosorb 102 was examined separately, but with a similar
testing protocol as employed for ethylene glycol, paraffin oil on
Chromosorb A and silica gel (Thomas, 1973; Thomas and Seiber, 1974).
The trapping efficiency of this material was every bit as good as
the other two solids when tested with 11 representative compounds.

We have conducted some preliminary tests with these three
solids at much higher flow rates. The device employed was a commercial high volume sampler, to the neck of which were added the
solids in approximately the same quantities as employed in the 5
cm diameter filter cup (Figure 3). Pesticides were spiked by
streaking on the filter. Both the filter and solid were analyzed
following the sampling period. The few results obtained to date
in checking this system have been encouraging but point up several
deficiencies (Table 8). Compounds of the DDT, cyclodiene (except
aldrin), and aryl thiophosphate classes are trapped well at flow
rates to 1 m³/min. Some chlorinated insecticides of high volatility, lindane, α-BHC, and β-BHC, were trapped better by the more
adsorptive silica gel than the two organic collecting agents.
Easily oxidized organophosphates such as trithion and ethion were
difficult to determine on all three solids, as was aldrin.

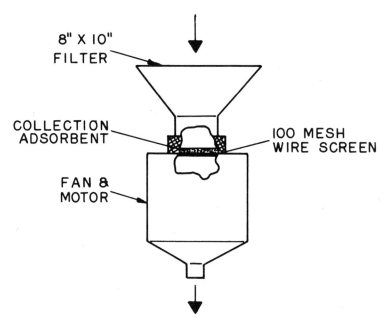

Figure 3.   High volume sampler modified for holding solid media.

The reagent background from the three solids did not differ greatly. The air background trapped by silica gel was somewhat higher than that from the other two solids. Detectability limits were estimated to be approximately 0.1 ng/m^3 for most of the compounds trapped by the three solids from outside air samples. Oxidative breakdown, measured by the conversion of parathion added directly to the solid to paraoxon after passage of 100 m^3 of air, was less with Chromosorb 102 (approximately 2%) than with paraffin oil on deactivated Chromosorb A or silica gel. These tests were carried out at 35°C to approximate the more severe conditions expected in midsummer; at milder temperatures the formation of paraoxon on all the solids was negligible.

Experiments are in progress to further test the concept of a simple pesticide trap capable of operation with commercial high volume air samplers. The nature of the trapping agent, its particle size and quantity, and the dimensions of the trap are variables. One design which seems appropriate would include a glass

TABLE 8

High Volume Sampling of Pesticide Vapors on Solid Sampling
Agents (100 m^3 Air Samples, 0.15 and 1.0 m^3/min)

	5% Paraffin Oil on Chrom A	Silica Gel	Chrom 102
	DDE	DDE	DDE
	DDT	DDT	DDT
Compounds	Hept. Epox.	Hept. Epox.	Hept. Epox.
trapped	Dieldrin	Dieldrin	Dieldrin
with	Endrin	Endrin	Me Parathion
50-100%	Ronnel	Ronnel	
efficiency	Me Parathion	Me Parathion	
	Et Parathion	Et Parathion	
	Diazinon	Diazinon	
		β-BHC	
		Lindane	
	Lindane	α-BHC	Lindane
20-50%	Ethion		α-BHC
	Malathion		
	Aldrin	Aldrin	Aldrin
0-20%	Trithion	Trithion	Trithion
	α-BHC	Ethion	
	β-BHC	Malathion	

fiber filter, for particulate collection, fitted on a container
holding the solid trapping agent adaptable to the top of the high
volume sampler. The adsorbent may be spread over an area of 100-
200 cm^2 to a depth of several centimeters. The resulting filter
pack could be prepared in the laboratory and transported intact
to and from the sampling site. A nagging detail remains to find
a meaningful way to spike air at these high flow rates to verify
the utility of such a device.

Bidleman and Olney (1974) described a system similar to this. Air was pulled at 0.4 to 0.8 m^3/min through a glass-fiber filter and 10 x 15 cm cylinder of polyurethane foam. Calibrating recoveries for Cl$_3$, Cl$_4$, and Cl$_5$ polychlorinated biphenyl vapors were better than 90%. This device was used to monitor chlorinated pesticide and PCB concentrations at Bermuda and nearby areas, and in Rhode Island. The results (Table 9) showed analytically significant levels of PCB, DDT, and chlordane in nearly all samples. Levels in air from Rhode Island contained considerably more of the compounds measured than those from Bermuda and the open sea. The levels of DDT measured at the oceanic sites were two orders of magnitude higher than those collected by the coated nylon net method for particulates of earlier studies (Risebrough et al., 1968; Seba and Prospero, 1971) suggesting that most of the chlorinated compounds exist in the atmosphere as vapors. This was supported by the small amount (less than 5%) of DDT found on the filter. The latter, however, may simply have been due to revolatilization of particulate DDT from the filter under the sampling conditions. At any rate, this polymeric medium appears as an excellent candidate for sampling; its efficiency towards other types of pesticides would be of great interest. The significant lowering of the detectability limit with this system when compared with ethylene glycol impingers reflects a major advantage offered by high volume sampling.

APPLICATIONS

Pesticide Transformation Products

Our attention to air sampling methods derived partly from an interest in determining the existence, if any, of pesticide transformation products in air. These could originate from two sources; evolved to the air after formation on or in condensed phases, and by breakdown, such as photoxidation, in the air. This explains our interest in the detail of breakdown in the sampling traps. It also explains our tendency to employ high volume air samplers, since any observed products would be expected to exist in only fractions of the parent compound concentration.

Trifluralin, a widely used and reasonably volatile preemergence dinitroaniline herbicide, was selected as a study model. Details of the vapor phase photodecomposition of trifluralin were reported elsewhere (Crosby et al., 1974b). Laboratory vapor-phase photoproducts of this chemical were trapped reasonably well from large volumes of air (120 m^3) using 5% paraffin oil on Chromosorb A in the high volume sampler (Table 10). The efficiencies were considerably higher for the more volatile compounds I-III at lower temperatures, but were reproducible at the more realistic 30°C even when

TABLE 9

Chlorinated Hydrocarbons in Marine and Continental Air From
High Volume Samples Collected Through Polyurethane Foam[a]

Collection dates (1973)	Location	Vol of air (m^3)	Conc. (ng/m^3)			
			PCB	p,p'-DDT	o,p'-DDT	Chlordane
2/12-2/13	Bermuda	1070	0.59	0.013	0.014	<0.005
6/4-6/5	33°20'N, 65°14'W	300	1.6	0.016	0.009	0.090
1/18-1/19	Providence	76	9.4	0.090	0.025	0.25

[a] Data selected from Bidleman and Olney (1974).

spiked at low levels (corresponding to 10 ng/m^3). These data,
incidentally, are typical of the temperature effects; they point
out a need for specifying temperatures when recovery data are presented.

Sampling was carried out just to the downwind side of a small
plot to which trifluralin had been applied at 1.5 lb/acre. Midday
air samples were taken after irrigation on the third day following
application, to promote vaporization. The results showed the un-
mistakable presence of breakdown products II-V in the high volume
air sample; since the same products were present in the soil,
however, it is difficult to ascertain whether they were formed in
the air or volatilized as such from the soil. We believe the former
contributed significantly and are designing experiments to verify
this.

Ambient Monitoring

A further impetus for developing reliable high volume sampling
methods comes from the recent renewal of interest in ambient air
monitoring for pesticides. The comprehensive program carried out
within EPA in 1970-71 (Yobs et al., 1972), the only one of its kind
to date, gives an idea of the types of pesticides frequently observed
(Table 11). DDT and its relatives, and dieldrin were the most common
residues; α-BHC, γ-BHC, and diazinon were also frequently encountered.
A number of other compounds (2,4-D esters, β-BHC, heptachlor, aldrin,

TABLE 10

Recoveries of Trifluralin and its Transformation Products
Using 5% Paraffin Oil on Chromosorb A
(120 m^3 Air Samples, 1 m^3/min)[a]

	14°C	30°C
Trifluralin (I)	68.9	39.9
Monopropyl (II)	67.3	36.0
Dealkylated (III)	67.2	49.3
Benzimidazole (IV)	--	85.0
Benzimidazole (V)	89.0	84.4

[a] Crosby et al., 1974b.

toxaphene, trifluralin, malathion, parathion) were more localized and seasonal in occurrence. Concentrations varied greatly, but were commonly in the 1-10 ng/m^3 range; the first value in this range reflects the working detectability limit of the ethylene glycol impinger system. Residues in lower concentrations, which could be valuable in correlating air levels with use patterns, were not generally measured. Furthermore, several important pesticide classes were not included in the cleanup methodology employed. Several goals for the new monitoring system thus emerge:

1. It should provide a working detectability limit of at least 0.1 ng/m^3 for most of the commonly encountered pesticides. A coupling of high volume sampling with efficient cleanup makes this goal appear within reach.

2. It should include a number of pesticides and related compounds not measured by the previous impinger system. This will entail development of new multiresidue fractionation methods in many cases.

3. It should allow at least spot checking for pesticide transformation products, particularly those with significant toxic potential. Organophosphorus "oxons," many phenols, and dioxins are examples.

4. It should provide a sufficient sample for confirmation. A sample from 1000 m^3 would easily be within reach of the common confirmation techniques of element-selective detection, derivation, and mass spectrometry, even when the concentrations are 0.1 ng/m^3 or less.

5. It should distinguish between vapor and particulate residues. Particle sizing, through use of cascade impactors, would be of even further value though it would multiply the number of analyses required per sample. This is an area in clear need of further research.

SUMMARY

The presence of pesticide residues in the air, a proven fact, suggests the prudence of continual analytical surveillance to assess the extent of contamination. Our knowledge of the nature, concentration, and form of parent compounds and their breakdown products of potential toxicological importance in the atmosphere is far from complete. Further definition will rest on the development and use of analytical methods capable of reliably measuring a wide range of chemicals at levels below 1 ng/m^3. For many applications, the ethylene glycol impinger offers a proven method of sampling, the key step in air analysis. It must be considered the standard by which the suitability of new methods may be judged. Since high volume air samplers are already in wide use for particulate analysis, a modification of these devices for pesticide determination is particularly attractive. Some details of design and calibration still remain to be worked out, but results obtained to date with inorganic, liquid-coated, and polymeric trapping media indicate that many of the analytical goals will be attained in the near future. It remains to be seen whether one single analytical system will suffice for all the chemicals of potential monitoring interest, in both vapor and particular forms. This seems a demanding but worthwhile objective for further study.

ACKNOWLEDGEMENT

Portions of this work were supported by National Science Foundation Grant GB-33723.

TABLE 11

EPA Division of Pesticide Community Studies
Monitoring Results, 1970-1971[a]

	1970	1971
Samples	882	968
Locations	30	45
Residues Occurring in >50% Samples	p,p'-DDT p,p'-DDE Dieldrin o,p'-DDT α-BHC γ-BHC Diazinon	p,p'-DDT p,p'-DDE Dieldrin o,p'-DDT α-BHC γ-BHC
Residues Occurring at all Sites	p,p'-DDT o,p'-DDT Dieldrin	p,p'-DDT
All-1	p,p'-DDE	Dieldrin

[a] Yobs et al., 1972

REFERENCES

Abbott, D.C., R.B. Harrison, J.O'G. Tatton, and J. Thomson. 1966. Organochlorine pesticides in the atmosphere. Nature 211:259.

Akesson, N.B., and W.E. Yates. 1964. Problems relating to application of agricultural chemicals and resulting drift residues. Ann. Rev. Entomol. 9:285.

Atkins, D.H.F., and A.E.J. Eggleton. 1970. Studies of atmospheric washout and deposition of gamma-BHC, dieldrin, and p,p-DDT using radio-labelled pesticides. AERE-R6528 1AEA/SM-142/32. Quoted by C.P. Lloyd-Jones, Atmos. Environ. 6:283, 1972.

Aue, W.A., and P.M. Teli. 1971. Sampling of air pollutants with support bonded chromatographic phases. J. Chromatogr. 62:15.

Bidleman, T.F., and C.E. Olney. 1974. Chlorinated hydrocarbons in the Sargasso Sea atmosphere and surface water. Science 183:516.

Bjorkland, J., B. Compton, and G. Zweig. September, 1970. Development of methods for collection and analysis of airborne pesticides. NAPCA Contract No. CPA 70-15 Report, Syracuse University Research Corporation.

Caro, J.H., A.W. Taylor, and E.R. Lemon. June, 1971. Measurement of pesticide concentrations in the air overlying a treated field. Proc. Intl. Sym. Ident. Measure. Environ. Pollutants, Ottawa, Ontario, pp. 72-77.

Cohen, J.M., and C. Pinkerton. 1966. Widespread translocation of pesticides in air transport and rain-out. In: Organic Pesticides in the Environment. R.F. Gould, editor, Washington, American Chemical Society, Adv. Chem. Ser. 60:163.

Compton, B., and J. Bjorkland. April, 1972. Design of a high-volume sampler for airborne pesticide collection. (021-Pesticide Chem.), 163rd Meeting, Division of Pesticide Chemistry, ACS, Boston, Massachusetts.

Crosby, D.G., K.W. Moilanen, C.J. Soderquist, and A.S. Wong. April, 1974a. Dynamic aspects of pesticide photodecomposition (004-Pesticide Chem), Division of Pesticide Chemistry, 167th Meeting, ACS, Los Angeles, California.

Crosby, D.G., K.W. Moilanen, J.N. Seiber, C.J. Soderquist, and J.E. Woodrow. April, 1974b. Photodecomposition of trifluralin vapor (070-Pesticide Chem.), Division of Pesticide Chemistry, 167th Meeting, ACS, Los Angeles, California.

Duggan, R.E., and G.Q. Lipscomb. 1969. Dietary intake of pesti-
     cide chemicals in the U.S. (II). June 1966 - April 1968.
     Pestic. Monit. J. 2:153.

Enos, H.F., J.F. Thompson, J.B. Mann, and R.F. Moseman. April,
     1972. Determination of pesticide residues in air. (022-
     pesticide Chem.), Division of Pesticide Chemistry, 163rd
     Meeting, ACS, Boston, Massachusetts.

Federal Environmental Pesticide Control Act of 1972. Public Law
     92-516, Washington, D.C., October, 1972, Section 20.

Freed, V.H., R. Haque, and D. Schmedding. 1972. Vaporization
     and environmental contamination by DDT. Chemosphere 1:61.

Gunther, F.A., W.E. Westlake, J.H. Barkley, W. Winterlin, and L.
     Langbehn. 1973. Establishing dislodgeable pesticide resi-
     dues on leaf surfaces. Bull. Environ. Contamin. Toxicol.
     9:243.

Hartley, G.S. 1969. Evaporation of pesticides. In: Pesticide
     Formulations Research. R.F. Gould, editor, Washington,
     American Chemical Society, Adv. Chem. Ser. 86:115.

Herzel, F., and E. Lahmann. 1973. Polyethylene-coated silica
     gel as a sorbent for organic pollutants in air. Z. Anal.
     Chem. 264:304.

Jegier, Z. 1969. Pesticide residues in the atmosphere. Ann. N.Y.
     Acad. Sci. 160:143.

Lewis, R.G. National Environmental Research Center, EPA, Research
     Triangle Park, North Carolina. Personal communication.

Lloyd-Jones, C.P. 1971. Evaporation of DDT. Nature 229:65.

Miles, J.W., L.E. Fetzer, and G.W. Pearce. 1970. Collection and
     determination of trace quantities of pesticides in air.
     Environ. Sci. Technol. 4:420.

Risebrough, R.W., R.J. Huggett, J.J. Griffin, and E.D. Goldberg.
     1968. Pesticides: Transatlantic movements in the northeast
     trades. Science 159:1233.

Seba, D.B., and J.M. Prospero. 1971. Pesticides in the lower
     atmosphere of the northern equatorial Atlantic Ocean. Atmos.
     Environ. 5:1043.

Shafik, M.T., and P.F. Mongan. April, 1972. Electron capture gas
    chromatography of picogram levels of methylcarbamate pesti-
    cides. (040-Pesticide Chem.), Division of Pesticide Chemistry,
    163rd Meeting, ACS, Boston, Massachusetts.

Sherma, J., and T.M. Shafik. 1974, in press. A multiclass, multi-
    residue analytical method for determining pesticide residues
    in air. Arch. Environ. Contam. and Toxicol.

Södergren, A. 1972. Chlorinated hydrocarbon residues in airborne
    fallout. Nature 236:395.

Spencer, W.F. 1970. Distribution of pesticides between soil,
    water, and air. In: Pesticides in the Soil, Int. Symp.
    Pestic. Soil, East Lansing, Michigan, Michigan State Univer-
    sity, pp. 120-128.

Spencer, W.F., and M.M. Cliath. 1972. Volatility of DDT and
    related compounds. J. Agr. Food Chem. 20:645.

Spencer, W.F., and M.M. Cliath. April, 1974. Vaporization of
    chemicals. (005-Pesticide Chem.), Division of Pesticide
    Chemistry, 167th Meeting, ACS, Los Angeles, California.

Stanley, C.W., J.E. Barney, M.R. Helton, and A.R. Yobs. 1971.
    Measurement of atmospheric levels of pesticides. Environ.
    Sci. Technol. 5:430.

Tabor, E.C. 1965. Pesticides in urban atmospheres. J. Air
    Pollut. Contr. Ass. 15:415.

Tessari, J.D., and D.L. Spencer. 1971. Air sampling for pesticides
    in the human environment. J. Ass. Off. Anal. Chem. 54:1376.

Thomas, T.C. 1973. Evaluation of Chromosorb 102 as a medium for
    trapping pesticides from air. M.S. Thesis, University of
    California, Davis, California.

Thomas, T.C., and J.N. Seiber. 1974, in press. Chromosorb 102,
    an efficient medium for trapping pesticides from air. Bull.
    Environ. Contam. Toxicol.

Van Middelem, C.H. 1971. Assay procedures for pesticide residues.
    In: Pesticides in the Environment. R. White-Stevens, editor,
    New York, Marcell Dekker, I(II):309.

Woodwell, G.M., P.P. Craig, and H.A. Johnson. 1971. DDT in the
    biosphere: where does it go? Science 174:1101.

Yobs, A.R., J.A. Hanan, B.L. Stevensen, J.J. Boland, and H.F. Enos. April, 1972. Levels of selected pesticides in ambient air of the United States. (024-Pesticide Chem.), Division of Pesticide Chemistry, 163rd Meeting, ACS, Boston, Massachusetts.

Yule, W.N., and A.F.W. Cole. 1969. Measurement of insecticide drift in forestry operations. Proc. Fourth Intl. Agricl. Aviation Congress, Kingston, Ontario, Canada.

Yule, W.M., A.F.W. Cole, and I. Hoffman. 1971. A survey for atmospheric contamination following forest spraying with fenitrothion. Bull. Environ. Contam. Toxicol. 6:289.

Zweig, G., G. Pirolla, and B. Compton. April, 1972. Monitoring and analysis of airborne pesticides. (025-Pesticide Chem.), Division of Pesticide Chemistry, 163rd Meeting, ACS, Boston, Massachusetts.

DYNAMIC ASPECTS OF PESTICIDE PHOTODECOMPOSITION

K.W. Moilanen, D.G. Crosby, C.J. Soderquist, and A.S. Wong

Department of Environmental Toxicology

University of California, Davis, California 95616

INTRODUCTION

The fact that sunlight wavelengths possess sufficient energy to effect photochemical transformations in organic molecules has created a great deal of interest in the environmental photochemistry of pesticides (Crosby and Li, 1969; Plimmer, 1970). Since those who conducted initial studies on this subject were faced with the problem of simulating natural conditions, many approaches were used--some apparently quite satisfactory and others with little or no relation to conditions encountered in nature. From our results and those reported in the literature, it has become increasingly evident that pesticide photochemistry is very dependent upon the environmental compartment in which the compound resides during irradiation.

Since pesticides may undergo leaching by water, volatilization into the atmosphere, adsorption to surfaces, and partitioning into organic films, their photochemical behavior in each of these compartments, listed in Table 1, is important if one is to successfully predict their environmental fate. The likelihood of exchange between compartments means the net result of the environmental photodecomposition of a pesticide may be a composite of its photochemical behavior in each.

The picture is made even more complex by the varying conditions within each compartment (Table 1). While this list certainly is not complete, it indicates several of the most important variables for each compartment, including pH, dissolved oxygen, natural photosensitizers, and nucleophiles in water; common air pollutants such as the oxides of nitrogen, ozone, hydrocarbons, and particulates

TABLE 1

Environmental Compartments and Their Important Variables

1. Water   2. Atmosphere

    a. pH   a. NO_x, O_3
    b. Dissolved oxygen   b. Hydrocarbons
    c. Natural photosensitizers   c. Particulates
    d. Nucleophiles

3. Surfaces   4. Organic Films

    a. pH   a. Dissolved oxygen
    b. Ionic forces   b. Natural photosensitizers
    c. Organic content
    d. Metal ions

___

in the atmosphere; pH, ionic forces, organic content, and metal ions
on surfaces; and dissolved oxygen and natural photosensitizers in
organic films.

THE WATER ENVIRONMENT

Laboratory Photolysis of 4-Chlorophenoxyacetic Acid

Figure 1 summarizes the photolysis of the growth regulator, 4-
chlorophenoxyacetic acid (4-CPA, I) (Crosby and Wong, 1973) and
illustrates the photolysis reactions commonly observed in aqueous
solution under laboratory conditions. The side chain is removed by
photooxidation to p-chlorophenyl formate (II) followed by hydrolysis
to p-chlorophenol (III), and this undergoes photonucleophilic hydrolysis to form hydroquinone (IV). Another pathway is reductive
dechlorination--replacement of the ring chlorine by hydrogen--to
produce phenoxyacetic acid (V) which undergoes two successive photooxidations to produce IV. The third pathway is initiated by photonucleophilic hydrolysis, followed by oxidation and hydrolysis to
again produce IV. The final step is rapid oxidation to humic acid
polymers.

Natural Photosensitization

Figure 2 shows the influence of sensitizers on the photolysis
rate of the rice herbicide, 2-methyl-4-chlorophenoxyacetic acid (VI)

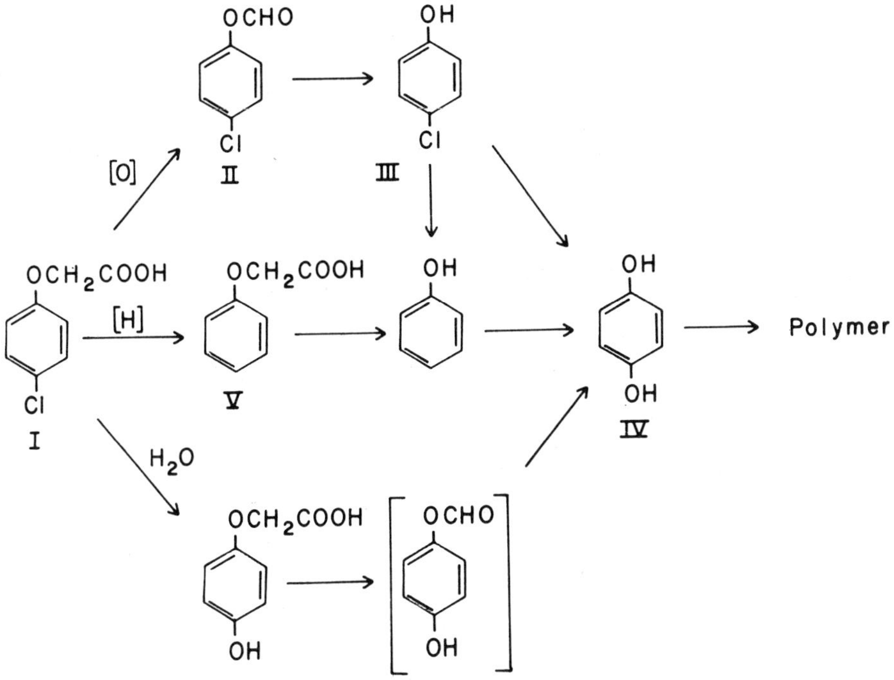

Figure 1.   Photolysis of 4-CPA in water (Crosby and Wong, 1973).

(Soderquist, 1973). Curve A depicts the photolysis rate in buffered
distilled water and indicates that 20 days of irradiation resulted
in 50% decomposition; Curve B, which represents the photolysis rate
in sterile (boiled) rice paddy water, shows that only 25% of VI
remained after only six days of irradiation. This example indicates
that agricultural waters contain compounds which can accelerate the
photolysis of organic compounds, and future investigations into the
environmental photochemistry of pesticides and related products will
have to consider the important feature of natural photosensitization.

Effect of pH

The hydrogen ion concentration can exert a strong influence on
the photolysis of a pesticide, as illustrated in Figure 3 , for the
herbicide trifluralin (2,6-dinitro-N,N-dipropyl-α,α,α-trifluoro-p-
toluidine (VII) (Crosby and Leitis, 1973). Under acidic conditions,

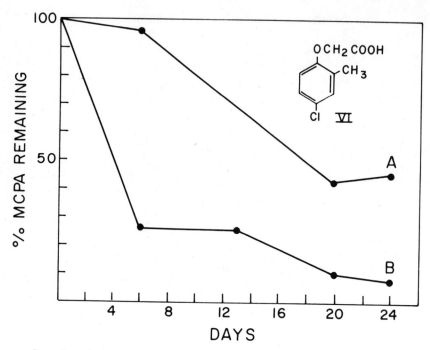

Figure 2.   Sensitized vs. unsensitized photolysis of MCPA (Soderquist and Crosby, in press).

photodecomposition was rapid and resulted primarily in polar products such as 2,3-dihydroxy-2-ethyl-7-nitro-1-propyl-5-trifluoro-
methylbenzimidazoline (X) (20% yield), 2-ethyl-7-nitro-5-trifluoro-
methylbenzimidazole-N-oxide (XI) (20% yield), and 2-amino-6-nitro-
α,α,α-trifluoro-p-toluidine (IX) (25% yield).   At pH 11, the product
profile was quite different, with 2-ethyl-7-nitro-5-trifluoromethyl-
benzimidazole (VIII), a compound not detected under acidic conditions, produced in 80% yield.   IX was produced in 8% yield, while
X and XI each were produced in only 5% yield at a photolysis rate
only 10% that at pH 5.5.

Oxygen Tension

Since photooxidations are commonplace (Crosby, 1972), the availability of dissolved oxygen (oxygen tension) is another important variable that must be taken into consideration in environmental photochemistry. Trifluralin again serves as a good illustration of how dissolved oxygen can alter a photolysis, as shown in Figure 4. Under aerobic conditions, both oxidation and cyclization occur, producing VII and 2,6-dinitro-N-propyl-α,α,α-trifluoro-p-toluidine (XII) which is rapidly converted to other products. Under anaerobic conditions, only cyclization occurs, resulting in 1-propyl-2-ethyl-7-nitro-5-trifluoromethylbenzimidazole (XIII) which accumulates since it is stable under these conditions. The photolysis rate is approximately ten times faster in the presence of air.

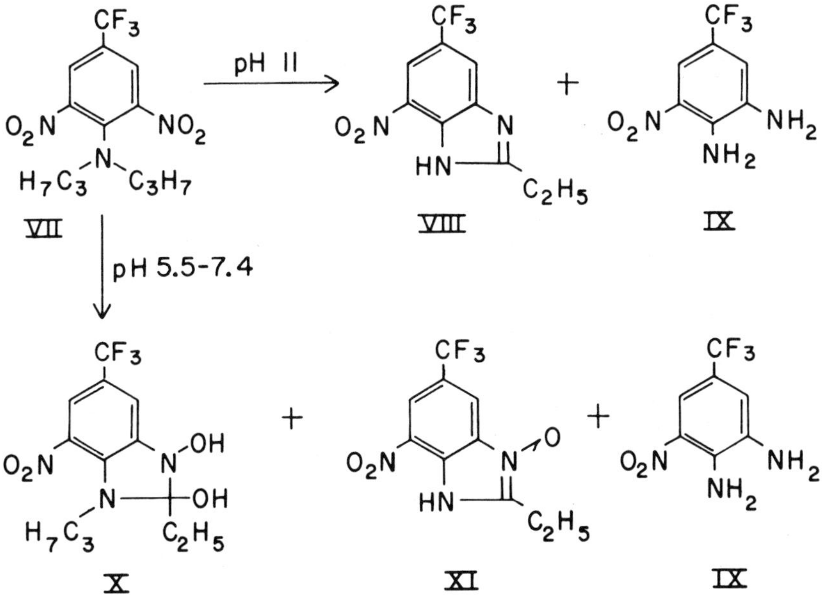

Figure 3. Photolysis of trifluralin under acidic and basic conditions (Crosby and Leitis, 1973).

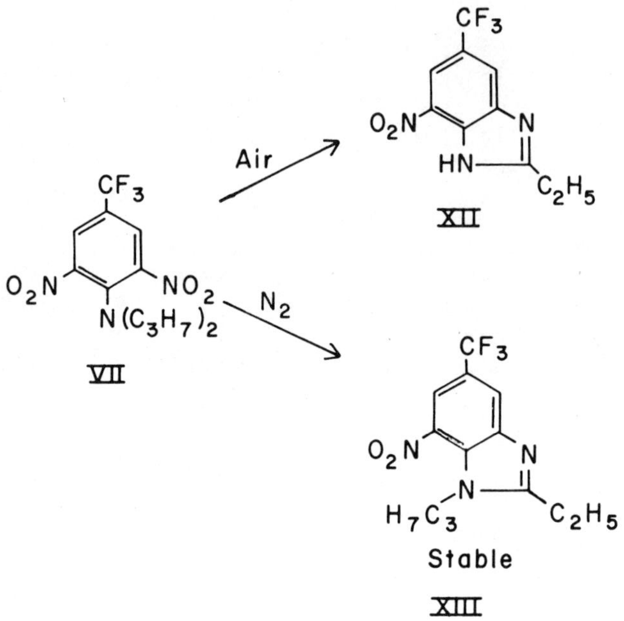

Figure 4. Photolysis of trifluralin under aerobic and anaerobic conditions.

Volatilization From Water Surfaces

Direct evaporation of pesticides from water surfaces appears to make a very important contribution to airborne residues as shown in Table 2 (Mackay and Wolkoff, 1973). Aldrin, with a solubility of 0.2 mg/1 and vapor pressure of 6 x 10^{-6} torr, has the surprisingly short half-life of 10.1 days. DDT, which is still less soluble and less volatile has an even shorter half-life. The PCB mixture, Aroclor 1254[R] and isooctane, with somewhat greater solubilities and much higher vapor pressures than any of the other compounds, and mercury, with its low solubility and moderate vapor pressure exhibit incredibly short residence times.

The rapid evaporative losses from water surfaces present an apparent contradiction, since these compounds have high molecular weights and relatively low vapor pressures. A factor which is often overlooked is the remarkably high activity coefficients of these

TABLE 2

Evaporation of Low-Volatility Compounds From Water Surfaces
(Mackay and Wolkoff, 1973)

Compound	Solubility mg/1	Vapor Pressure torr	Half-life
Aldrin	2×10^{-1}	6×10^{-6}	10.1 days
DDT	1.2×10^{-3}	1×10^{-7}	3.7 days
Aroclor 1254	1.2×10^{-2}	7.71×10^{-5}	1.2 min.
2,2,4-Trimethylpentane	2.44	49.3	17.9 min.
Mercury	3×10^{-2}	1.3×10^{-3}	4.1 sec.

compounds in water which cause unexpectedly high equilibrium partial
pressures and thus high rates of evaporation. While the data in
Table 2 indicate that evaporation from rivers and lakes may serve
as an important means of transport of pesticides to the atmosphere,
the actual losses under environmental conditions probably would be
slower due to rate-limiting diffusion and desorption.

AIR ENVIRONMENT

Vapor-Phase Photodecomposition

The evaporation or volatilization of pesticides leads to consideration of another environmental compartment--the atmosphere.
This compartment, which once was considered relatively unimportant
to the fate of pesticides, is now recognized as a major transport
medium for them (Risebrough et al., 1968; Woodwell et al., 1971).
Since environmental photooxidations (photooxygenations) are commonplace in condensed phases (Crosby, 1972), they also would be expected to occur readily in the atmosphere due to plentiful oxidant,
sunlight, and catalytic surfaces. Unimolecular reactions such as
photoisomerization and intramolecular condensation which are independent of collision with other molecular species also are to be
expected.

To conduct well-controlled laboratory studies, a reactor was required which would simulate atmospheric conditions as closely as possible. Figure 5 shows a schematic diagram of such a vapor-phase photoreactor (Crosby and Moilanen, 1974) in which a collimated beam of ultraviolet (UV) light is delivered through the center of a 72-liter spherical reaction chamber containing the pesticide vapor and is then absorbed to prevent reflection back onto the reactor wall. The spherical chamber and collimated light provide very little opportunity for wall reactions, a highly desirable feature in vapor-phase photochemical investigations. The all-glass reactor is air-tight and can accommodate substances ranging from low volatility solids to gases such as methyl bromide.

Figure 6 illustrates both the photooxidation and photoisomerization expected in the vapor-phase (Crosby and Moilanen, 1974); aldrin (1,2,3,4,10,10-hexachloro-1,4,4a,5,8,8a-hexahydro-1,4-endo, exo-5,8-dimethanonaphthalene, XIV) vapor underwent epoxidation to dieldrin (1,2,3,4,10,10-hexachloro-6,7-epoxy-1,4,4a,5,6,7,8,8a-octahydro-1,4-endo,exo-5,8-dimethanonaphthalene, XV) and isomerization to photoaldrin (1,1,2,3,3a,7a-hexachloro-2,3,3a,3b,4,6a,7,7a-octahydro-2,4,7-methano-1H-cyclopenta[a]pentalene, XVI). Dieldrin vapor was isomerized to photodieldrin (1,1,2,3,3a,7a,-hexachloro-5,6-epoxidecahydro-2,4,7-methano-1H-cyclopenta[a]pentalene, XVII), as also reported by Nagl et al. (1970), and XVI was epoxidized to XVII. These reactions were not affected by the intentional irradiation of the reactor wall, indicating that they were authentic vapor-phase transformations. Prolonged irradiation of photodieldrin vapor resulted in no further reaction and indicated that, under reactor conditions, it represents the stable end-product of aldrin photodecomposition.

The ease of oxidation of aldrin and photoaldrin vapor via epoxidation of a double bond indicates that this type of reaction may apply generally for chlorinated olefins in the atmosphere. This premise is supported by previous work on the photochemistry of tettachloroethylene (XVIII) (Dickinson and Leermakers, 1932; Dickinson and Carrico, 1934; Frankel et al., 1957) in which irradiation by sunlight in the presence of oxygen resulted in the intermediate formation of tetrachloroethylene oxide (XIX) which was rapidly converted to trichloroacetyl chloride (XX) and phosgene (XXI) as shown in Figure 7.

Figure 8 depicts the vapor-phase photochemistry of p,p'-DDT [1,1,1-trichloro-2,2-*bis*-(p-chlorophenyl) ethane, XXII] (Moilanen and Crosby, 1973) and a mechanism for the formation of the observed 4,4'-dichlorobenzophenone (XXV). The initial reaction, dehydrochlorination to p,p'-DDE [1,1-dichloro-2,2-*bis*-(p-chlorophenyl) ethylene, XXIII] occurs readily, indicating that the photochemistry of DDT actually represents DDE photochemistry. The next step is

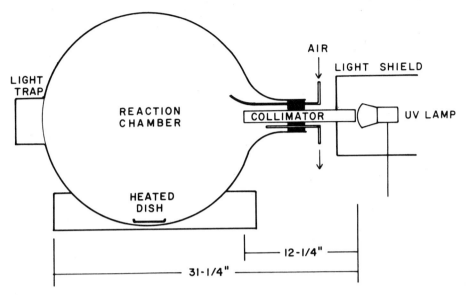

Figure 5. Schematic diagram of the vapor-phase photoreactor (Crosby and Moilanen, 1974).

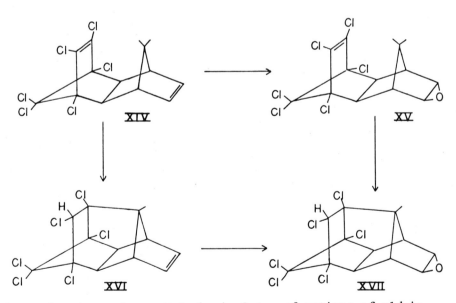

Figure 6. Vapor-phase photochemical transforations of aldrin, dieldrin, and photoaldrin (Crosby and Moilanen, 1974).

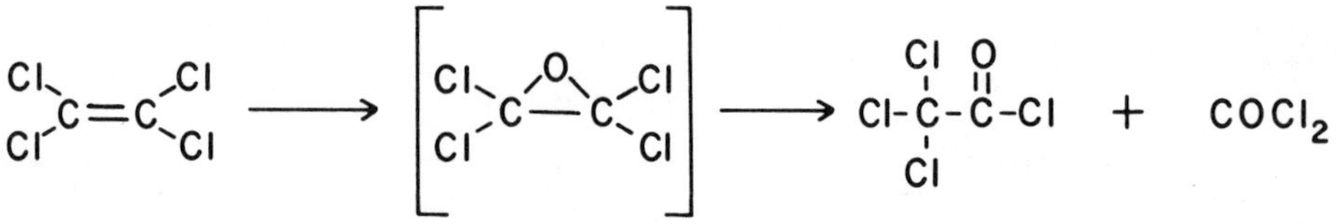

XVIII XIX XX XXI

Figure 7. Vapor-phase photochemical transformations of tetrachloroethylene.

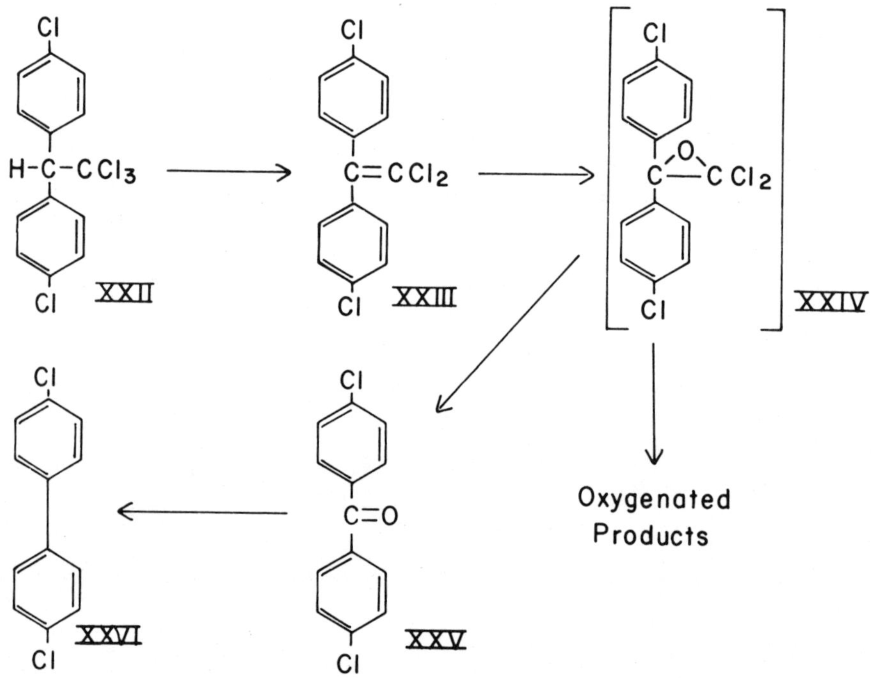

Figure 8. Vapor-phase photochemical transformations of *p,p'*-DDT.

the proposed photo-epoxidation of DDE to produce 1,1-dichloro-2,2-
bis-(*p*-chlorophenyl) ethylene oxide (XXIV) (Crosby, Leffingwell,
and Moilanen, 1973). While this intermediate has not as yet been
isolated, it successfully accounts for the formation of XXV, since
epoxides are known to rearrange to products analogous to XXV.
Irradiation of XXV vapor resulted in slow decarbonylation to form
a dichlorobiphenyl (XXVI) which was stable toward further vapor-
phase irradiation indicating that, under reactor conditions, it
represents a stable end-product of DDT photodecomposition.

Several investigations of the environmental photochemistry of
parathion [*O,O*-diethyl *O*-(*p*-nitrophenyl) phosphorothioate, XXVII] in
condensed phases have been reported (Cook and Pugh, 1957; Frawley
et al., 1958; Koivistoinen, 1962; Koivistoinen and Merilainen, 1962;
Grunwell and Erickson, 1973) and indicate that it was readily converted to paraoxon [*O,O*-diethyl *O*-(*p*-nitrophenyl) phosphate, XXVIII]
and other products. Consequently, parathion vapor was irradiated
to determine if the expected photooxidation would occur under vapor-
phase conditions. Figure 9 indicates that parathion vapor was not
photooxidized to paraoxon; in fact, parathion was stable toward UV
light and did not photodecompose even after extended irradiation.

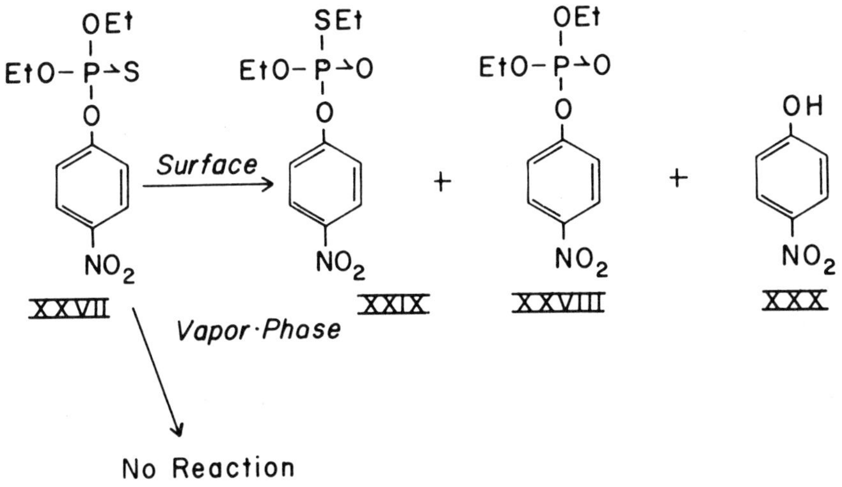

Figure 9. Photoproducts from parathion adsorbed on dust particles.

THE SURFACE ENVIRONMENT

This rather surprising result suggested that degradation of parathion might take place in a different environmental compartment, perhaps on solid surfaces. Since a logical extension of vapor-phase studies involves the photochemical reactions of pesticides adsorbed on airborne particulates, parathion was applied to fine road-dust and exposed to the same light used in the vapor-phase experiment; conversion to paraoxon (XXVIII), O,S-diethyl O-(p-nitrophenyl) phosphate (XXIX), and p-nitrophenol (XXX) was rapid (Figure 9). The conversion of parathion to decomposition products commonly found in field samples illustrates the importance which particulate surfaces may play in the environmental photochemistry of pesticides.

THE ORGANIC FILM ENVIRONMENT

The presence of thin films of oil on the surface of natural bodies of water and the likelihood that the particulates suspended in air are coated with organic matter leads to consideration of yet another environmental compartment, organic films. Superficially, one might assume that photochemical reactions in thin films would be identical to those in bulk organic solvents. However, further consideration of the physical situation represented by an organic film--a very thin layer of organic solvent with reducing ability, the abundant exposure to sunlight and oxygen, and the opportunity for extraction and concentration of pesticides and natural photosensitizers from the adjacent water--suggests that reactions in films could differ greatly from those expected.

Thin films could exert a strong influence upon transport processes at the air/water and air/surface interfaces. In some cases, films could act as insulators preventing direct contact between air and water, thus reducing evaporation rates for certain pesticides. Also, compounds which tend to undergo photochemical reduction might photolyze at a much faster rate in films while oxidations such as the conversion of aldrin to dieldrin might be slowed considerably by lack of oxygen solubility.

Figure 10 shows the photolysis of 3,3',4,4'-tetrachlorobiphenyl (XXXI) in hexane solution (Ruzo et al., 1972), where the only reaction was reductive dechlorination to photo-products having fewer chlorine atoms. The resistance of monochlorobiphenyl (XXXIV) to further photolysis indicates that it may be an end product of PCB photolysis in thin films.

Figure 11 shows the contrasting photochemical behavior of 2,2',5,5'-tetrachlorobiphenyl (XXXV) when irradiated in aqueous solution (Crosby, Moilanen, and Wong, 1973). Both reductive

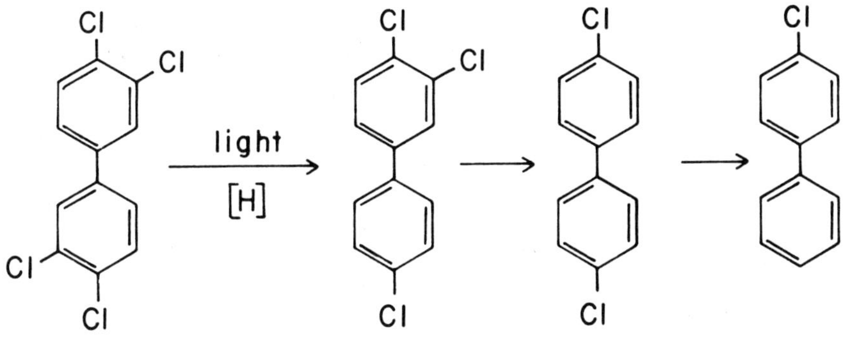

XXXI XXXII XXXIII XXXIV

Figure 10. Photolysis of 3,3',4,4'-tetrachlorobiphenyl in hexane (Ruzo, Zabik, and Shuetz, 1972).

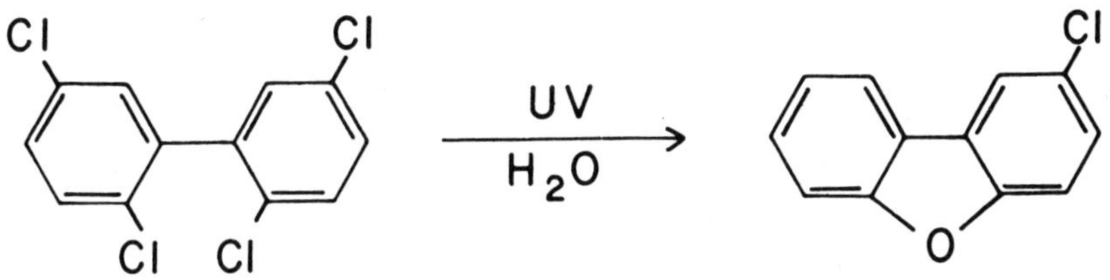

XXXV XXXVI

Figure 11. Photochemical generation of 2-chlorodibenzofuran from 2,2',5,5'-tetrachlorobiphenyl (Crosby, Moilanen, and Wong, 1973).

dechlorination and photonucleophilic hydrolysis occurred to produce 2-chlorodibenzofuran (XXXVI). XXXVI is formed when the chlorine in the 2-position is displaced by hydroxide ion and the oxygen atom of the resulting phenoxide ion displaces the neighboring chlorine atom.

CONCLUSION

These examples illustrate that pesticides may be expected to undergo photochemical reactions readily under a wide variety of environmental conditions. The photolytic rate and nature of the photoproducts are dependent upon the chemical properties of both the parent compound and the environmental compartment in which photolysis takes place. The probability of intercompartment movement of pesticides and their photoproducts suggests that photodegradation in the environment is a continual, dynamic process. The variables which exist within each compartment (Table 1) often exert a crucial influence on the photolysis of a pesticide and further emphasize the dynamic nature of environmental photochemistry.

Many compounds which are, by themselves, resistant to photolysis rapidly break down in the presence of naturally occurring sensitizers, oxidants, hydrogen-donors, etc., and disappear from the environment much faster than one would predict from laboratory studies. Further, the rapid evaporation of aldrin and DDT from water indicates that many pesticides of relatively low vapor pressure may readily volatilize into the atmosphere where movement and photochemical reactions largely determine their fate. There, the existence of particulate matter such as dust and smoke provides an opportunity for the adsorption of pesticide molecules on these surfaces; once adsorbed, photochemical behavior may be markedly altered, as in the case of parathion. The widespread presence of thin organic films on solid surfaces and on the surfaces of natural bodies of water provides still a different type of medium for photochemical transformation.

The net result of the environmental photochemistry of a pesticide can be assumed to be a composite of its photochemical behavior in each of the environmental compartments. Considering the variety and availability of reagents and the powerful energy of the UV component of sunlight, the frequently rapid and unaccounted "dissipation" of pesticides comes into clearer focus. Dynamic transport offering continual opportunity for the described types of transformations indeed may virtually preclude any exact balance-sheet for pesticides and many other chemicals in the environment.

ACKNOWLEDGMENT

This work was supported, in part, by a grant from the National Science Foundation (GB-33723).

REFERENCES

Cook, J.W., and N.D. Pugh. 1957. A quantitative study of cholinesterase-inhibiting decomposition products of parathion formed by ultraviolet light. J. Assoc. Off. Agr. Chem. 41:399.

Crosby, D.G. 1972. Environmental photooxidation of pesticides. Degradation of Synthetic Organic Molecules in the Biosphere. National Academy of Sciences, Washington, DC, 260.

Crosby, D.G., J.T. Leffingwell, and K.W. Moilanen. 1973. Transformations of environmental contaminants by light. The Third International Symposium on Chemical and Toxicological Aspects of Environmental Quality, Tokyo, Japan, November 19-22.

Crosby, D.G., and E. Leitis. 1973. The photodecomposition of trifluralin in water. Bull. Environ. Contam. Toxicol. 10:237.

Crosby, D.G., and M.Y. Li. 1969. In, Degradation of Herbicides. P.C. Kearney and D.D. Kaufman (eds). Dekker, New York, 321.

Crosby, D.G., and K.W. Moilanen. 1974. Vapor-phase photodecomposition of aldrin and dieldrin. Arch. Environ. Contam. Toxicol. 2:62.

Crosby, D.G., K.W. Moilanen, and A.S. Wong. 1973. Environmental generation and degradation of dibenzodioxins and dibenzofurans. Environ. Health Perspectives 5:259.

Crosby, D.G., and A.S. Wong. 1973. Photodecomposition of p-chlorophenoxyacetic acid. J. Agr. Food Chem. 21:1049.

Dickinson, R.G., and J.L. Carrico. 1934. The photochlorination and the chlorine-sensitized photo-oxidation of gaseous tetrachloroethylene. J. Amer. Chem. Soc. 56:1473.

Dickinson, R.G., and J.A. Leermakers. 1932. The chlorine-sensitized photo-oxidation of tetrachloroethylene in carbon tetrachloride solution. J. Amer. Chem. Soc. 54:3852.

Frankel, D.M., C.E. Johnson, and H.M. Pitt. 1957. Preparation and properties of tetrachloroethylene oxide. J. Org. Chem. 22:1119.

Frawley, J.P., J.W. Cook, J.R. Blake, and O.G. Fitzhugh. 1958. Effect of light on chemical and biological properties of parathion. J. Agr. Food Chem. 6:28.

Grunwell, J.R., and R.H. Erickson. 1973. Photolysis of parathion (O,O-diethyl-O-(4-nitrophenyl)thiophosphate). New products. J. Agr. Food Chem. 21:929.

Koivistoinen, P. 1962. The effect of sunlight on the disappearance of parathion residues. Acta. Agr. Scand. 12:285.

Koivistoinen, P., and M. Merilainen. 1962. Paper chromatographic studies on the effect of ultraviolet light on parathion and its derivatives. Acta. Agr. Scand. 12:267.

Mackay, D., and A.W. Wolkoff. 1973. Rate of evaporation of low-solubility contaminants from water bodies to atmosphere. Environ. Sci. Technol. 7:611.

Moilanen, K.W., and D.G. Crosby. 1973. Vapor-phase photodecomposition of p,p'-DDT and its relatives. Paper No. 21, Division of Pesticide Chemistry, 165th Meeting, American Chemical Society, Dallas, Texas.

Plimmer, J.R. 1970. The photochemistry of halogenated herbicides. Residue Reviews 33:47.

Risebrough, R.W., R.J. Huggett, J.J. Griffin, and E.D. Goldberg. 1968. Pesticides: Transatlantic movements in the northeast trades. Science 159:1233.

Ruzo, L.O., M.J. Zabik, and R.D. Schuetz. 1972. Polychlorinated biphenyls: Photolysis of 3,3',4,4'-tetrachlorobiphenyl and 4,4'-dichlorobiphenyl in solution. Bull. Environ. Contam. Toxicol. 8:217.

Soderquist, C.J. 1973. Dissipation of MCPA in a rice field. M.S. Thesis, University of California, Davis, California.

Woodwell, G.M., P.P. Craig, and H.A. Johnson. 1971. DDT in the biosphere: where does it go? Science 174:1101.

VAPORIZATION OF CHEMICALS*

W.F. Spencer and M.M. Cliath

USDA, Agricultural Research Service, University of

California, Riverside, California 92502

Vaporization from soil, plant, and water surfaces and atmospheric transport are important in the dissipation and movement of many chemicals. For some time, it has been recognized that highly volatile pesticides are lost from treated surfaces mainly by vaporization. However, only recently has it been established that appreciable quantities of even the so-called nonvolatile pesticides, such as the organochlorine insecticides, move into the atmosphere by vaporization. Regardless of inherent volatility, the same physical and chemical principles govern rates of vaporization and movement of chemicals from treated surfaces.

Potential volatility of a chemical is related to its vapor pressure, but actual vaporization rate will depend on environmental conditions and all factors that modify or attenuate the effective vapor pressure of the chemical. The effective vapor pressure may differ from the vapor pressure of the chemical itself as a result of such factors as adsorption on soil or other surfaces, solution in the soil water, solution in oily or waxy surfaces of leaves and/or fruits, penetration into plant surfaces, or retreat into deeper soil capillary spaces. Vaporization of a pesticide from soil is controlled by many variables, including temperature, soil properties, soil water content, and nature of the pesticide--particularly its solubility and its degree of adsorption (Hamaker, 1972; Spencer et al., 1973). Similar variables will control vaporization of pesticides from other adsorbing surfaces, although

---

* Contribution from the Agricultural Research Service, USDA, in cooperation with the California Agricultural Experiment Station, Riverside.

very little quantitative data are available on the relationship
between vapor pressure of a pesticide and its adsorption or penetration into leaf or fruit surfaces.

This paper will deal mainly with factors affecting vaporization rates, mechanisms of vaporization, measurement of vaporization
losses, and models to predict vaporization rates, with particular
emphasis on vaporization of chemicals from soils.

MECHANISMS AND FACTORS THAT INFLUENCE VAPORIZATION RATES

Vaporization from surface deposits is dependent only on vapor
pressure of the pesticide and its rate of movement away from the
evaporating surface. However, when a pesticide is mixed into the
soil, loss by vaporization involves desorption of the pesticide
from the soil, movement to the soil surface, and vaporization into
the atmosphere. Consequently, soil-incorporated pesticides vaporize
at a greatly reduced rate dependent upon the effective vapor pressure of the pesticide within the soil and its rate of movement to
the surface. Mechanisms and factors affecting vaporization of
chemicals will be discussed from the standpoint of: 1) those that
affect movement away from the evaporating surfaces; 2) those that
affect vapor pressure of the pesticide or its partitioning between
the soil, water, and air in and above the soil; and 3) those that
affect movement of the pesticide to the evaporating surface.

Pesticide Movement Away From the Evaporating Surface

When a substance vaporizes into air, its rate of vaporization
is determined solely by its vapor pressure (or vapor density) and
its rate of diffusion through the air closely surrounding the substance. Close to the evaporating surface there is relatively no
movement of air, and the vaporized substance is transported from
the surface through this stagnant air layer only by molecular
diffusion. Molecular diffusion coefficients of organic compounds
in air are inversely proportional to the square root of molecular
weight (Hartley, 1969). The actual rate of mass transfer by diffusion will be proportional to the diffusion coefficient and to
the vapor density, but since the vapor density is proportional to
the vapor pressure (p) times molecular weight (M), the rate of loss
under standard conditions will be proportional to $p(M)^{1/2}$. Diffusion away from the surface will thus be related to the vapor pressure of the pesticide, its molecular weight and, of course, temperature which influences both vapor pressure and the diffusion
process itself. Hartley (1969) demonstrated that one can reasonably
predict vaporization rate on the basis of vapor pressure, if vapor
pressure and vaporization rate for a model compound are known, by
use of the equation:

$$F_b = \frac{P_b (M_b)^{1/2}}{P_a (M_a)^{1/2}} \cdot F_a \qquad (1)$$

where F is vapor flux, P is vapor pressure, M is molecular weight, and the subscripts (a) and (b) refer to the model compound and the vaporizing chemical, respectively.

Rate of air flow over the evaporating surface affects the rate of evaporation, since the depth of the stagnant air layer will depend on the air flow rate. Air movement also affects vaporization by continuously replacing the air around the evaporating surface and by mixing of air composition due to turbulence. Surface geometry can alter air flow effects through its influence on air turbulence.

Vaporization of chemicals from isolated particles or droplets, such as may occur in drifting sprays, is controlled, not only by the vapor density and rate of movement of the chemical away from the surface, but also by the rate at which heat can be supplied to the evaporating surface for the latent heat of change of state.

In the case of surface deposits, vaporization of a chemical from a complete surface of constant area is independent of the depth of the chemical layer, and under constant conditions the loss rate is constant until so little substance remains that it can no longer cover the surface. Consequently, the rate of loss per unit time will be approximately the same whether the complete surface deposit is 10 $\mu g/cm^2$ or 1,000 $\mu g/cm^2$. To avoid misinterpretation of data, volatilization losses should not be expressed as percent of the deposit lost per unit time unless information is included also on deposit density. If a pesticide deposit is applied to a glass surface at a high concentration, the percent loss may be low per unit time even if vaporization rates on a unit area basis are relatively high. For example, Fleck (1944) reported loss of DDT from a glass surface at 45 C of only 6.6% in 37 days. This led him to conclude that vaporization of DDT from spray deposits will be too slow to be of any importance. The 6.6% loss in 37 days is equivalent to 2.27 $\mu g/cm^2/day$. If a DDT residue of approximately 10 $\mu g/cm^2$ were present on a leaf surface, the same loss rate reported by Fleck would result in complete loss of residue after four and one-half days. Consequently, even though the reported percent loss by vaporization is relatively low, the loss rate in $\mu g/cm^2$ is sufficiently high to account for a high percentage of DDT lost from many plant surfaces. Clearly, data on evaporation from surface deposits, such as deposits on glass plates, should always be expressed in absolute terms. The densities of pesticide deposits remaining on treated plant surfaces are usually much lower than densities used in most studies with glass plates. Consequently, data on loss rates from glass plates should be interpreted carefully unless comparisons with other surfaces are made on the basis of loss rates per unit area.

In water-pesticide systems, vaporization of water and the pesticide is a diffusion-controlled process with no mutual influence (Hamaker, 1972; Spencer et al., 1973; Hartley, 1969). However, the high affinity of some chemicals for the periphery of the liquid in aqueous suspensions causes some pesticides to concentrate at the water surface, resulting in vaporization rates greater than would be predicted from the average concentration of the pesticide in the water suspension. Such was reported by Bowman et al. (1959) and Acree et al. (1963) for DDT. This enhanced vaporization, due to the heterogeneous distribution of DDT in aqueous suspension, was attributed to codistillation by Acree et al. (1963). Codistillation implies a carrier-distillation process which modifies the volatility of the material being distilled by lowering the temperature at which diffusive transfer in the gas phase changes to bulk flow when the combined vapor pressures equal or exceed the ambient pressure. This type of bulk-flow process in the gas phase does not occur in vaporization of pesticides and water at temperatures far below the boiling point of water.

Pesticide Vapor Density

Every chemical has a characteristic saturation vapor density or vapor pressure which varies with temperature. Vapor pressure also is influenced by adsorption on surfaces, such as soils, and by solution in water or other liquids, such as plant oils and waxes. Spencer and Cliath (1969) developed a method for measuring vapor pressure of pesticides in soil which makes it possible to evaluate directly the effects of soil and environmental factors on potential volatility of individual pesticides. As expected, vapor pressures of pesticides are greatly decreased by their interactions with soils--mainly due to adsorption (Spencer and Cliath, 1969, 1970, 1972, and in press; Spencer et al., 1969). How much adsorption reduces the vapor pressure of a pesticide in soil depends mainly on the nature of the pesticide, pesticide concentration in soil, soil water content, and soil properties such as organic matter and clay content.

Adsorption reduces the chemical activity, or fugacity, below that of the pure compound. This is then reflected in changes in vapor pressure of the chemical. Bailey and White (1964, 1970) and White and Mortland (1970) reported factors such as soil or colloid type, physiochemical nature of the pesticide, soil reaction, temperature, nature of the saturating cation on the exchange sites, and nature of the pesticide formulation directly influenced the

adsorption-desorption of pesticides by soil systems. For weakly
polar or non-ionic pesticides, the amount of soil organic matter
is the most important soil factor for increasing adsorption and,
consequently, for decreasing vapor pressure or potential volatility of a pesticide added to the soil. With more polar or ionic
molecules, clay minerals play an increasingly important role in
adsorption and volatility affects. Most of the more volatile
pesticides are only weakly polar or non-ionic; thus, their adsorption by soils is closely related to organic matter content. For
example, Spencer (1970) reported that dieldrin vapor pressure in
five soils varied inversely with soil organic matter content. Likewise, Guenzi and Beard (1970) found that the initial volatilization
rates of DDT and lindane were inversely related to the organic
matter content and surface area of four soils.

Soil pesticide concentration. Concentration of the pesticide
chemical at the soil surface will determine whether, and to what
extent, adsorption by the soil will reduce vapor pressure or
potential volatility. Soil surface applications, or pesticides
falling on the soil surface from foliage applications, can result
in relatively high concentrations at the evaporating surface. For
example, a 1-kg/ha application of pesticide is equivalent to 10
$\mu g/cm^2$, or approximately 150 ppm in the surface 0.5 mm of soil.

The vapor pressure of weakly polar pesticides in soil increases
greatly with increases in pesticide concentration and reaches saturated vapor densities equal to that of the pesticide without soil
at relatively low soil pesticide concentrations. In moist Gila
silt loam, saturation vapor densities for dieldrin (Spencer et al.,
1969), lindane (Spencer and Cliath, 1970), p,p'-DDT (Spencer and
Cliath, 1972), o,p'-DDT (Spencer and Cliath, 1972), and trifluralin
(Spencer and Cliath, in press) were reached at soil concentrations
of 25, 55, 15, 39, and 73 $\mu g/g$, respectively. The relatively low
soil concentrations needed for a saturated vapor in moist soil
indicate that weakly polar pesticides, such as these, may have
relatively high vaporization losses, especially when applied to a
moist soil surface. Incorporation of the pesticide into the soil
would, of course, decrease the concentration at the evaporating
surface, thereby reducing volatilization.

Soil water content. Pesticides vaporize much more rapidly from
wet than from dry soils (Guenzi and Beard, 1970; Fang et al., 1961;
Harris and Lichtenstein, 1961; Deming, 1963; Kearney et al., 1964;
Bowman et al., 1965; Gray and Weierich, 1965; Parochetti and Warren,
1966; Willis et al., 1971; Igue et al., 1972). Measurements of
vapor pressures of pesticides in soil at various water contents
conclusively demonstrated that greater vaporization from wet than
from dry soils is due mainly to an increased vapor pressure resulting from displacement of the pesticide from the soil surface by
water (Spencer et al., 1969). Vapor pressure of dieldrin in soil

dropped to a very low value when the water content was decreased
below that equivalent to one molecular layer. However, vapor
pressure increased to its original maximum value upon rewetting the
dry soil, indicating that the drying effect was reversible. Similar
marked decreases in vapor pressure in dry soils were reported for
lindane (Spencer and Cliath, 1970), DDT compounds (Spencer and
Cliath, 1972), and trifluralin (Spencer and Cliath, in press). This
water content-vapor density effect on vaporization was confirmed by
Igue et al., 1972) and Guenzi and Beard (1970) with lindane, DDT,
and dieldrin.

Increased volatility due to displacement by water would undoubtedly occur with all weakly polar compounds for which water
can compete for adsorption sites. Weakly adsorbed pesticides and
water also compete for adsorption sites on other surfaces, and the
effect of humidity on their vaporization from surfaces such as
glass, metal, plant foliage, etc., can be accounted for by the
higher humidity furnishing water molecules to displace the pesticide from the solid surface.

Temperature. Temperature influences vaporization rate mainly
through its effect on vapor pressure of the pesticide. Vapor pressures of pesticides follow the usual reciprocal-temperature relations, $\text{Log}_{10} p = A - B/T$, where A and B are constants, and p and T
refer to vapor pressure and absolute temperature, respectively.
Spencer and Cliath (1969) reported that, as temperature increased,
the increase in vapor pressure of dieldrin was the same whether
the dieldrin was in soil or without soil. The heat of vaporization
(a measure of the change in vapor pressure with temperatures since
$\Delta Hv = 4.57 B$) of dieldrin with or without soil was 23.6 kcal/mol.
Pesticides more strongly adsorbed than dieldrin may exhibit heats
of vaporization in soil different from those of the pure chemicals.

Temperature may also influence vaporization rate of soil-
incorporated pesticides through its effect on movement of the pesticide to the surface by diffusion or by mass flow in the evaporating water, or through its effect on the soil water adsorption-
desorption equilibrium. For these effects also, increases in temperature are usually associated with increases in vaporization
rates. The only exception would be where rapid soil drying from
the higher temperature restricted vaporization more than temperature increased it.

Pesticide Movement Toward the Evaporating Surface

When a pesticide is mixed into the soil, the initial vaporization rate will depend on the soil concentration-vapor density
relationships at the soil surface. But vaporization rate decreases
sharply as the concentration at the soil surface is depleted and

soon becomes dependent upon rate of movement of the pesticide to
the soil surface (Spencer and Cliath, in press; Spencer, 1970;
Spencer and Cliath, 1973; Farmer et al., 1972; Farmer et al., 1973).
Diffusion and mass flow in evaporating water are the two general
mechanisms whereby pesticides move to the evaporating surface. In
the absence of evaporating water, vaporization rate depends on rate
of movement of the pesticide to the soil surface by diffusion only.
When water evaporates from the soil surface, the suction gradient
produced results in an appreciable upward movement of water to replace that evaporated, and any pesticide in the soil solution moves
toward the surface by mass flow with the evaporating water. Usually
both mechanisms operate together in the field where water and pesticides vaporize at the same time.

   Diffusion. Whenever a concentration gradient is present, a
chemical diffuses from the area of higher concentration to the area
of lower concentration. Vaporization depletes the soil surface of
pesticide. This, in turn, causes additional pesticide to diffuse
upward along the concentration gradient to replace that vaporized.
Hamaker (1972) and Letey and Farmer (1974) recently discussed diffusion of pesticides in soils. In general, the same factors control diffusion rate of a chemical that control its vapor pressure;
i.e., temperature and variables affecting adsorption, such as pesticide concentration and the soil water, organic matter, and clay
contents. In addition, soil bulk density influences rate of diffusion (Farmer et al., 1973).

   In the absence of water movement to the evaporating surface,
diffusion equations can be used to predict changes in pesticide
concentration within the soil and the loss rate of a pesticide at
the soil surface by vaporization, if diffusion coefficients for the
chemical in the soil are known. Mayer et al. (in press) developed
diffusion models that predicted volatilization rates in close
agreement with rates measured by Igue et al. (1972) and Spencer
and Cliath (1973). Figure 1 illustrates the comparison between
volatilization rates measured by Spencer and Cliath and those predicted by a diffusion model of Mayer et al. Vaporization of soil-
incorporated dieldrin was measured by passing an air stream, at
100% relative humidity to prevent loss of water, over the soil
surface. For the comparison illustrated in Figure 1, Model II,
described by Mayer et al., was used:

$$F = DC_0/(\pi Dt)^{1/2} \tag{2}$$

where F = pesticide flux through the surface, D = diffusion coefficient in soil, C_0 = initial pesticide concentration, and t = time.
The model assumes that the pesticide concentration at the soil
surface was maintained at zero and that the soil column was of infinite depth. Mayer et al. (in press) developed four other mathematical models for predicting volatilization under various boundary

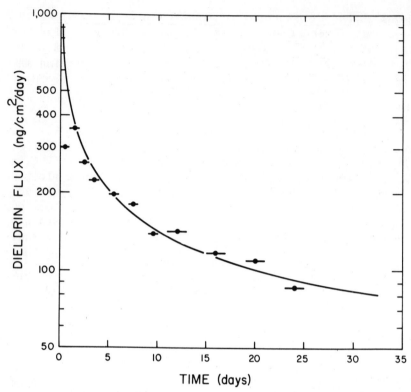

Figure 1. The comparison between calculated values (solid curve)
and experimental values (horizontal lines) for dieldrin flux. The
calculated values were obtained using Model II, assuming a value
of 2.3 mm^2/week for diffusion coefficient of dieldrin (Mayer et al.,
in press).

conditions based on pesticide movement to the soil surface by diffusion only. All models require a knowledge of diffusion coefficients for the particular pesticide in the soil being used. A
knowledge of pesticide adsorption isotherms is required in two of
the models where pesticide concentrations in the air, as well as in
the soil, are important.

Mass flow in water moving to the surface. Water moving upward
to replace that evaporated will carry pesticides in solution toward
the evaporating surface. This has been referred to as the "wick
effect" by Hartley (1969). Spencer and Cliath (1973) demonstrated
that movement of pesticides toward the soil surface by mass flow
in evaporating water can accelerate pesticide volatilization.

Vaporization of soil-incorporated lindane and dieldrin was measured
under conditions whereby the soil surface remained moist while the
relative humidity (R.H.) of the air passing over the soil was varied
to regulate water loss from the soil. Figure 2, from Spencer and
Cliath (1973), shows lindane vaporization rate as related to R.H.
of the air passing over the soil surface. Water loss accelerated
pesticide vaporization rate, but only after the soil surface had
been depleted of the pesticide, indicating that the enhancement was
due to the wick effect and not codistillation. During the first
four hours with R.H. alternating between 100 and 50% at one-half
hour intervals, vaporization rates were higher at 100% R.H.--when
no water was evaporating. This indicates that codistillation was
not operating and evaporation of water itself did not enhance lindane vaporization rate. Vaporization rate decreased rapidly as
the surface became depleted of lindane, and the evaporation of
water at 50% R.H. resulted in a higher lindane flux than at 100%
R.H. because of mass flow of lindane to the surface in the evaporating water. During subsequent periods of continuous 100% R.H.,
lindane flux rates decreased to relatively low values. When the
R.H. was decreased, causing water to evaporate and move water
upward, vaporization rate increased to points B_2, C, and D for
relative humidities of 50, 15, and 0%, respectively.

Data obtained for both dieldrin and lindane (Spencer and Cliath,
1973) indicate that vaporization rates due to the mass flow effect
can be estimated from the relationship:

$$F_p = F_w \cdot c \qquad\qquad\qquad\qquad (3)$$

where F_p is the pesticide flux, F_w is the water flux, and c is the
concentration of pesticide in the soil water. To use this relationship to predict pesticide loss due to mass flow, data for the
rate of water loss are needed as well as adsorption coefficients
relating pesticide concentrations in the soil to soil solution
concentrations. The magnitude of the wick effect will depend upon
the adsorption characteristics and water solubility of the pesticide
and other factors affecting partitioning between the water, air, and
solid phases in the soil. With lindane, the flux due to mass flow
ranged from 356 to 703 ng/cm^2/day, equivalent to 18 to 71% of the
total lindane flux. Dieldrin flux due to the mass flow effect
ranged from 9.9 to 23.6 ng/cm^2/day, equivalent to 3 to 33% of the
total dieldrin flux.

Equation (3) may be useful for predicting long-term loss rates,
but may be difficult to use for predicting short-term vaporization
losses. At very low relative humidities, dieldrin accumulated at
the soil surface, resulting in accelerated vaporization when the
surface was again moistened by exposure to a R.H. of 100% (Spencer
and Cliath, 1973). Consequently, difficulties in predicting short-

term vaporization losses may arise because of rapid drying of the
soil surface which could result in a reduced vapor density at the
surface, temporarily reducing pesticide vaporization rate instead
of increasing it, and causing the pesticide to accumulate at the
soil surface. Results with dieldrin indicate that such accumulated
pesticides will vaporize when the soil is again moistened. If so,
vaporization due to mass flow over the long term will be related to
water loss and can probably be predicted from total water evaporated,
if concentrations of pesticide in the evaporating water can be estimated. For a more detailed discussion of mechanisms and factors
that influence vaporization rates of pesticides see Hamaker (1972)
and Spencer et al. (1973).

MAGNITUDE OF VAPORIZATION RATES

Surface Deposits

Vaporization rate of a surface chemical residue is proportional
to the gross area of the deposit and can be predicted within reasonable limits with equation (1), if vaporization rate of a model compound is known. Spencer et al. (1973) used the equation to calculate
potential vaporization rates of several pesticides using measured
dieldrin losses (Spencer and Cliath, 1973) as the vaporization flux
of the model compounds (F_a in equation (1)). Estimated vaporization
rates at 30C from surface deposits of dieldrin, p,p'-DDT, lindane,
parathion, and trifluralin were 2.23, 0.14, 17.9, 23.8, and 50.4
$\mu g/cm^2$/day, respectively. These vaporization rates are those to be
expected with flow rates, geometry, temperature, etc., used in the
laboratory apparatus described by Spencer and Cliath (1973). The
exact relation between vaporization rates measured in the laboratory
and rates measured in the field is not known. However, rates of
vaporization in the field probably would be higher than those above
because of the higher rate of air exchange under most field conditions.

Many pesticides have vapor pressures in the range between those
of lindane (1.28×10^{-4} at 30C) and dieldrin (1.0×10^{-5} at 30C)
and vaporization rates, such as the above, probably account for the
very rapid disappearance of deposits on foliage and fruits reported
by several workers (Ebeling, 1963; Gunther, 1969). For example, a
typical 10 $\mu g/cm^2$ deposit of an insecticide, with a vapor pressure
similar to dieldrin and a volatilization rate of 2.23 $\mu g/cm^2$/day,
would disappear in somewhat less than five days. A laboratory
confirmation of the relatively high vaporization rates from foliage
was obtained by measuring vapor losses of parathion residues on
detached citrus leaves from sprayed groves (Spencer and Cliath,
unpublished). At a residue concentration of approximately 1.5 $\mu g/$

cm^2 of leaf surface, parathion vaporized at the rate of 0.73 µg/cm^2/ day at 30C when air was passed over the leaves at 3.4 liters per minute.

Adsorption or penetration into the fruit or leaf surface and solubilization in plant waxes and oils would retard vaporization, but the initial deposit should vaporize at a rate closely resembling the pure material, because only a small fraction of the applied pesticide would be in direct contact with the leaf or fruit surfaces. Residues that do penetrate leaf or fruit surfaces and become solubilized in plant waxes or oils disappear very slowly as compared with rates of disappearance of the initial deposits (Gunther, 1969; Jeppson and Gunther, 1970).

Soil-Incorporated Pesticides

The simplified relation expressed by equation (1) cannot be used to predict long-term vaporization of soil-incorporated pesticides where vaporization decreases rapidly with time and levels off at a much lower rate dependent on movement of the pesticide to the surface. A comparison of data for soil-incorporated lindane in Figure 2 with the estimate from surface deposits illustrates the great difference in potential vaporization rates from surface deposits of chemicals compared with those from soil-incorporated materials. At 30C, the estimated vaporization rate from a surface deposit of lindane was calculated to be 23.8 µg/cm^2/day, compared with a loss rate from 10 µg/g lindane in Gila silt loam of 1.96 µg/cm^2/day, initially (first 24 hours), and 0.26 µg/cm^2/day after 26 days.

Likewise, trifluralin vaporized much more rapidly when applied on the surface of Gila silt loam (Figure 3) than when mixed into the soil (Figure 4) (Spencer and Cliath, in press). The maximum volatilization rate from surface applications was approximately 40 µg/cm^2/day (4 kg/ha/day) and the 1- and 2-kg/ha applications to wet soils were almost completely lost by vaporization in less than 24 hours. When 14 kg/ha was incorporated to 10 cm (10 µg/g), the loss rate during the first 24 hours was only 0.52 µg/cm^2/day (0.052 kg/ha/day) or 0.36% of that applied. Because of the higher rate of air exchange expected in the field than over the laboratory columns, rates of vaporization from surface applications in the field probably would be higher than those in Figure 3. The gas over the soil columns was exchanged, on the average, once every 3 seconds, equivalent to a linear flow rate of only 0.039 mi/hr. This is relatively low compared to windspeeds encountered in most agricultural areas. However, since the loss of soil-incorporated pesticides are controlled more by their rate of movement to the soil surface than by the rate of air exchange over the soil, volatilization rates of soil-incorporated trifluralin in the field probably would be similar to those reported in Figure 4.

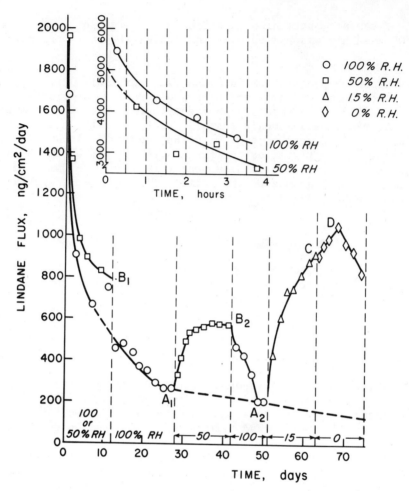

Figure 2. Lindane volatilization rate as related to relative humidity of the N_2 gas passing over the soil surface at 30°C with 10 ppm lindane mixed in Gila silt loam at a water suction of 50 millibars (Spencer and Cliath, 1973).

High rates of vaporization, similar to those observed from surface applications of trifluralin, undoubtedly occur with many pesticides, but at differing rates dependent on their vapor pressures. Such high rates of vaporization contributes to the short-term effectiveness of many insecticides, to the low-use efficiency of some herbicides and to the environmental contamination with pesticides.

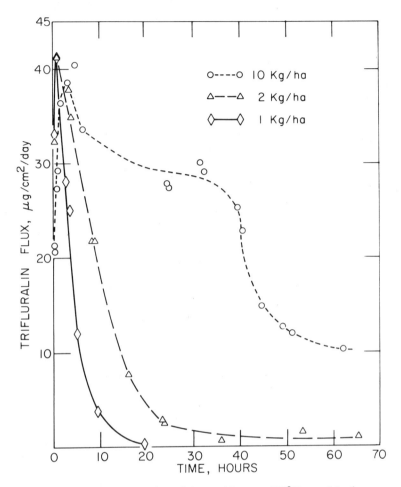

Figure 3. Volatilization of trifluralin at 30°C applied on a wet soil surface of Gila silt loam at 10, 2, and 1 kg/ha (Spencer and Cliath, in press).

An accurate evaluation of pesticide vaporization losses from field soils involves the use of equipment to trap the emanating vapors, meteorological instrumentation, and soil measurements as described by Parmele et al. (1972) and used by Caro et al. (1971) to measure dieldrin volatilization rates from field applications. Caro et al. periodically measured the flux rate into the air of dieldrin and heptachlor applied at the rate of 5.6 kg/ha and incorporated into the soil by disking, followed by planting of corn. The vaporized pesticides were collected at several heights by extraction

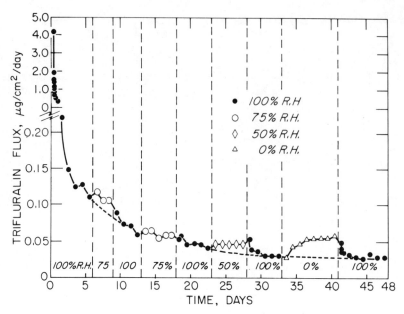

Figure 4. Trifluralin volatilization rate as related to relative humidity of the N_2 gas passing over the soil surface at 30°C with 10 μg/g trifluralin mixed in Gila silt loam at a water suction of 50 mbars (expanded scale first day only) (Spencer and Cliath, in press).

of the air in hexylene glycol. Concentrations were determined by gas chromatography of the extracts. Vaporization rates were estimated from pesticide concentration gradients in the atmosphere by use of vertical diffusivity coefficients calculated from simultaneous water vapor losses as described by Parmele et al. (1972). Pesticide vapor losses for the entire growing season were estimated from flux data for several sampling days. Dieldrin loss for the season was 250 g/ha, or 4.5% of the applied material, and loss of heptachlor was 346 g/ha, or 6.2% of that applied. This extremely complex study is the only known measurement of actual volatilization rates from soil-incorporated pesticides conducted under normal farming conditions while a crop was growing and conclusively demonstrates that vaporization of insecticides is a major pathway of loss from the soil under normal management practices.

Atmospheric concentrations of pesticides above treated fields were measured by Willis et al. (1969, 1971) and by Cliath and Spencer (1972) and Spencer et al. (1974). Even though actual volatilization rates were not estimated, these measurements have been of considerable value in confirming the relatively high loss rates of pesticides under field conditions.

Pesticide degradation products also volatilize and in some cases this could be a more important pathway for dissipation of soil-incorporated pesticides than direct loss of applied material. For example, Cliath and Spencer (1972) found that degradation products of DDT and lindane are much more volatile than the parent compounds. Field measurements of atmospheric concentrations of various DDT compounds indicated that 66% of the "total DDT" in the atmosphere above a field containing residual DDT was p,p'-DDE (Cliath and Spencer, 1972; Spencer et al., 1974).

SUMMARY

Evidence is presented to demonstrate the importance of vaporization and vapor phase movement in dissipation of pesticides from plant, water, and soil surfaces. Vaporization is discussed from the standpoint of the mechanisms involved, factors influencing rates of vaporization, and progress in developing models for predicting vapor flux under various conditions. Many variables affect vaporization rate, but most of them do so through their effects on vapor pressure, rate of movement away from the evaporating surface, or rate of movement to the evaporating surface.

Vaporization of chemicals from surface deposits is dependent only on vapor pressure of the chemical and its rate of movement away from the evaporating surface. Soil-incorporated chemicals or chemicals that have penetrated fruit or foliage vaporize at a greatly reduced rate dependent upon the effective vapor pressure of the chemical within the soil, or within the plant and its rate of movement to the evaporating surface. The effective vapor pressure differs from the vapor pressure of the chemical itself owing to adsorption on soil or other surfaces, solution in soil water, or solution in the oil or wax of leaves or fruit. The magnitude of the adsorption effect, or reduction of the vapor pressure of a chemical in soil, is dependent mainly on the nature of the chemical, soil chemical concentration, soil water content, and soil properties such as organic matter and clay content.

Soil water content effects are especially important. Pesticides vaporize faster from wet soils than from dry soils, primarily because water increases the vapor pressure of the pesticide by competing for adsorption sites. However, mass flow in water moving to the surface for evaporation can also contribute to the greater pesticide vaporization from wet than from dry soils. Although considerable progress has been made in quantifying factors affecting vaporization and in developing mathematical models for predicting vaporization rates, much additional work is needed to more accurately predict vaporization losses under conditions in which pesticides are actually being used on growing crops.

REFERENCES

Acree, F., Jr., M. Beroza, and M.C. Bowman. 1963. Codistillation of DDT with water. J. Agric. Food Chem. 11:278.

Bailey, G.W., and J.L. White. 1964. Review of adsorption and desorption of organic pesticides by soil colloids with implications concerning pesticide bioactivity. J. Agric. Food Chem. 21:324.

Bailey, G.W., and J.L. White. 1970. Factors influencing the adsorption, desorption, and movement of pesticides in soil. Res. Rev. 32:29.

Bowman, M.C., F. Acree, Jr., C.H. Schmidt, and M. Beroza. 1959. Fate of DDT in larvicide suspensions. J. Econ. Entomol. 52: 1038.

Bowman, M.C., M.S. Schechter, and R.L. Carter. 1965. Behavior of chlorinated insecticides in a broad spectrum of soil types. J. Agric. Food Chem. 13:360.

Caro, J.H., A.W. Taylor, and E.R. Lemon. 1971. Measurement of pesticide concentrations in air overlying a treated field. International Symp. on Identification and Measurement of Environmental Pollutants, Ottawa, Canada, pp. 72.

Cliath, M.M., and W.F. Spencer. 1972. Dissipation of pesticides from soil by volatilization of degradation products. I. Lindane and DDT. Environ. Sci. Tech. 6:910.

Deming, J.M. 1963. Determination of volatility losses of C^{14}-CDAA from soil samples. Weeds 11:91.

Ebeling, W. 1963. Analysis of the basic processes involved in the deposition, degradation, persistence, and effectiveness of pesticides. Res. Rev. 3:35.

Fang, S.C., P. Theisen, and V.H. Freed. 1961. Effects of water evaporation, temperature, and rates of application on the retention of ethyl-N,N-di-n-propylthiocarbamate in various soils. Weeds 9:569.

Farmer, W.J., K. Igue, and W.F. Spencer. 1973. Effect of bulk density on the diffusion and volatilization of dieldrin from soil. J. Environ. Qual. 2:107.

Farmer, W.J., K. Igue, W.F. Spencer, and J.P. Martin. 1972. Volatility of organochlorine insecticides from soil. I. Effect of concentration, temperature, air flow rate, and vapor pressure. Soil Sci. Soc. Amer. Proc. 36:443.

Fleck, E.E.   1944.   Rate of evaporation of DDT.   J. Econ. Entomol.
    37:853.

Gray, R.A., and A.J. Weierich.   1965.   Factors affecting the vapor
    loss of EPTC from soils.   Weeds 13:141.

Guenzi, W.D., and W.E. Beard.   1970.   Volatilization of lindane and
    DDT from soils.   Soil Sci. Soc. Amer. Proc. 34:443.

Gunther, F.A.   1969.   Insecticide residues in California citrus
    fruits and products.   Res. Rev. 28:1.

Hamaker, J.W.   1972.   Diffusion and volatilization.   In Organic
    Chemicals in the Soil Environment, Dekker, New York, pp. 341.

Harris, C.R., and E.P. Lichtenstein.   1961.   Factors affecting the
    volatilization of insecticidal residues from soils.   J. Econ.
    Entomol. 54:1038.

Hartley, G.S.   1969.   Evaporation of pesticides.   In Pesticidal
    Formulations Research, Physical and Colloidal Chemical
    Aspects.   Adv. Chem. Series 86:115.

Igue, K., W.J. Farmer, W.F. Spencer, and J.P. Martin.   1972.
    Volatility of organochlorine insecticides from soil.   II.
    Effect of relative humidity and soil water content on dieldrin
    volatility.   Soil Sci. Soc. Amer. Proc. 36:447.

Jeppson, L.R., and F.A. Gunther.   1970.   Acaricide residues on
    citrus foliage and fruits and their biological significance.
    Res. Rev. 33:101.

Kearney, P.C., T.J. Sheets, and J.W. Smith.   1964.   Volatility of
    seven s-triazines.   Weeds 12:83.

Letey, J., and W.J. Farmer.   1974.   Movement of pesticides in soil.
    In Pesticides and Their Effects on Soil and Water.   Amer. Soc.
    Agron., Madison, Wisconsin.

Mayer, R., W.J. Farmer, and J. Letey.   1974, in press.   Models for
    predicting pesticide volatilization of soil-applied pesticides.
    Soil Sci. Soc. Amer. Proc.

Parmele, L.H., E.R. Lemon, and A.W. Taylor.   1972.   Micrometeoro-
    logical measurement of pesticide vapor flux from bare soil and
    corn under field conditions.   Water, Air, and Soil Pollution
    1:433.

Parochetti, J.V., and G.F. Warren. 1966. Vapor losses of IPC and
    CIPC. Weeds 14:281.

Spencer, W.F. 1970. Distribution of pesticides between soil,
    water, and air. In Pesticides in the Soil: Ecology, Degrada-
    tion, and Movement. Michigan State University, East Lansing,
    Michigan, pp. 120.

Spencer, W.F., and M.M. Cliath. 1969. Vapor density of dieldrin.
    Environ. Sci. Tech. 3:670.

Spencer, W.F., and M.M. Cliath. 1970. Desorption of lindane from
    soil as related to vapor density. Soil Sci. Soc. Amer. Proc.
    34:574.

Spencer, W.F., and M.M. Cliath. 1972. Volatility of DDT and
    related compounds. J. Agric. Food Chem. 20:645.

Spencer, W.F., and M.M. Cliath. 1973. Pesticide volatilization
    as related to water loss from soil. J. Environ. Qual. 2:284.

Spencer, W.F., and M.M. Cliath. 1974, in press. Factors affecting
    vapor loss of trifluralin from soil. J. Agric. Food Chem.

Spencer, W.F., and M.M. Cliath. Unpublished data.

Spencer, W.F., M.M. Cliath, and W.J. Farmer. 1969. Vapor density
    of soil-applied dieldrin as related to soil-water content,
    temperature, and dieldrin concentration. Soil Sci. Soc. Amer.
    Proc. 33:509.

Spencer, W.F., M.M. Cliath, W.J. Farmer, and R.A. Shepherd. 1974.
    Volatility of DDT residues in soil as affected by flooding and
    organic matter applications. J. Environ. Qual. 3:126.

Spencer, W.F., W.J. Farmer, and M.M. Cliath. 1973. Pesticide
    volatilization. Res. Rev. 49:1.

White, J.L., and M.M. Mortland. 1970. Pesticide retention by clay
    minerals. In Pesticides in the Soil: Ecology, Degradation,
    and Movement. Michigan State University, East Lansing,
    Michigan, pp. 95.

Willis, G.H., J.F. Parr, R.I. Papendick, and S. Smith. 1969. A
    system for monitoring atmospheric concentrations of field-
    applied pesticides. Pest. Monit. J. 3:172.

Willis, G.H., J.F. Parr, and S. Smith. 1971. Volatilization of
    soil-applied DDT and DDD from flooded and nonflooded plots.
    Pest. Monit. J. 4:204.

MODELING OF ATMOSPHERIC BEHAVIOR: A SUBMODEL OF THE DYNAMICS OF PESTICIDES

W. Brian Crews, John W. Brewer*, and Timothy J. Petersen

Department of Mechanical Engineering

University of California, Davis, California 95616

This paper is a review of standard methods for estimating the diffusion of substances in the atmosphere. Such estimates may be used to analyze a portion of the dynamics of pesticides in the environment.

Other submodels of pesticide dynamics are discussed elsewhere in the symposium and deal with such important phenomena as:

1. The sources of pesticide (quantity, temporal and geographical distribution, physical and chemical nature);

2. The sinks or scavenging mechanisms (physical, chemical, biological);

3. Transformation by biological and physical agents;

4. Hydrological diffusion and transport;

5. Bioaccumulation;

6. Environmental response (impact).

Submodels of all of these effects must be combined with an atmospheric diffusion submodel in order to provide a useful analysis of the environmental dynamics of pesticides (Figure 1).

The review begins with a discussion of time and geographical scales used in the interpretation of calculations of atmospheric

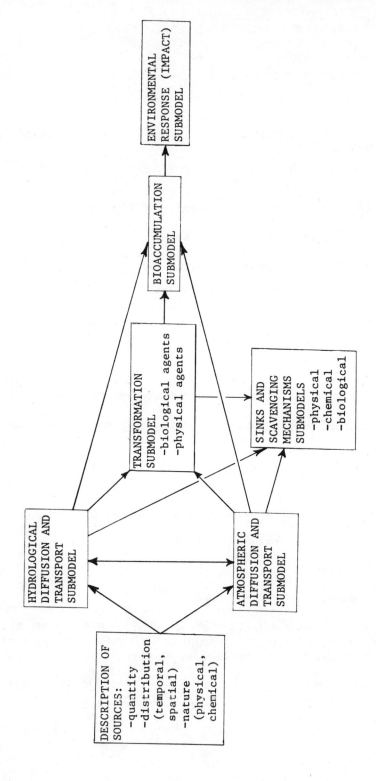

Figure 1. A systems model of the environmental dynamics of a pesticide. This paper is a review of methods for assembling the atmospheric diffusion submodel.

diffusion and transport. A brief description is then provided of
atmospheric diffusion modeling methods and the review concludes
with a list of references.

THE QUESTION OF SCALES

All mathematical models are simplified descriptions of a
phenomenon. Usually another set of simplifications and approxima-
tions must be made in order to obtain numerical results from the
mathematical model. In the case of models of the transport of
pesticides, these simplifications will place restrictions on the
time and geographical scales for which the simulation results may
be considered accurate. A trade-off will exist between the degrees
of resolution and the costs of model development and simulation.
A basic principle of diffusion modeling is that the degrees of
resolution of time and geographical scales be consistent with the
goals of the analyst.

The principle proposed above seems self-evident and not worthy
of emphasis. However, inattention to the principle is not unpre-
cedented. The difficulty arises because the user of a particular
modeling technique is not always aware of the entire set of simplifications involved and, therefore, of the constraints on the degrees
of resolution. The main body of this review is devoted to a de-
lineation of the simplifications involved in the commonly applied
techniques for modeling the transport of substances in the atmosphere.

An attempt is now made to illustrate the value of the proposed
principle. Crews (1973) recently demonstrated that rough estimates
of the exposure of urban population to lead aerosol can be obtained
from a modest amount of meteorological and source data. The method
is based on the steady state solution of the species continuity
(diffusion) equation. The method may well have application in the
study of pesticides. The nature of the data and simplifications
used by Crews requires that his simulation results be interpreted
as monthly and city-wide averages; hence, his technique can be used
to estimate the chronic exposure of the environment to lead aerosol, but would not be useful in the analysis of acute response to
short-term exposure. To properly assess the maximum exposure of
specific demographic groups, a much finer resolution of time and
geographical scales must be made.

A more precise description of "scale" and "degree of resolution" is attempted. The degree of resolution is used here as
synonymous with averaging time: the length of time used to smooth
widely fluctuating instantaneous measurements. An averaging time
may be a property of the measurement instrument or may be purposely
introduced to eliminate the effects of short-term turbulent

fluctuations. The geographical scale is roughly the extent of the
region being studied by the analyst. The time scale is taken to
be the range of time between the averaging time and the period of
the largest turbulent fluctuations within the geographical region.
Thus, a relation exists between geographical and time scales.

The authors propose the scale classification illustrated in
Figure 2. This classification is based upon a liberal interpretation and extension of the meteorological classification implied by
Monin (1972). (The authors' "international" class corresponds to
the "global" scale of Monin.)

The classification scheme is used as an aid to the selection
of atmospheric diffusion models for estimating exposure levels.
To illustrate the use of the scheme, suppose that a pesticide is
only moderately volatile and highly reactive. It might be decided
that the crucial environmental problem is that of instantaneous
exposure of farm workers. The authors would delegate such a problem
to the micro or meso class. On the other hand, a mildly toxic pesticide which is highly persistent and volatile would be studied for
its long-term impact on, say, the marine biota. Obviously, such a
problem should be associated with one of the global classifications
(and the long-term persistence property would be the determining
factor for the selection of a specific global class).

Since scavenging mechanisms and all toxic responses are difficult to anticipate, it is expected that the classification of a
given pesticide problem will change with time. As new scavenging
mechanisms are discovered, the classification will move to smaller
time scale classifications. As new toxic responses are discovered,
the classification will probably tend to move in the direction of
finer time resolution.

THE CLASSIFICATION OF TRANSPORT MODELS--
EXAMPLES FROM THE LITERATURE

A brief compilation of atmospheric diffusion modeling techniques will be attempted for most of the classes illustrated in
Figure 2*. Atmospheric diffusion modeling may be divided into three
types of approaches. The two best developed approaches are "probabilistic" and "physical" models. A third is the "statistical"
(regression) model. The probabilistic approach began with Sutton
(1932) and has led to the development of the very useful "plume"
and "puff" type models. Physical models are based upon an analysis
of the mechanisms of mass conservation and result in "continuity
equations" which may be solved for pollutant concentration. An
example of a continuity equation is the time averaged form:

---

* This is certainly not the first attempt of this nature; for
  example, see Munn and Bolin (1971).

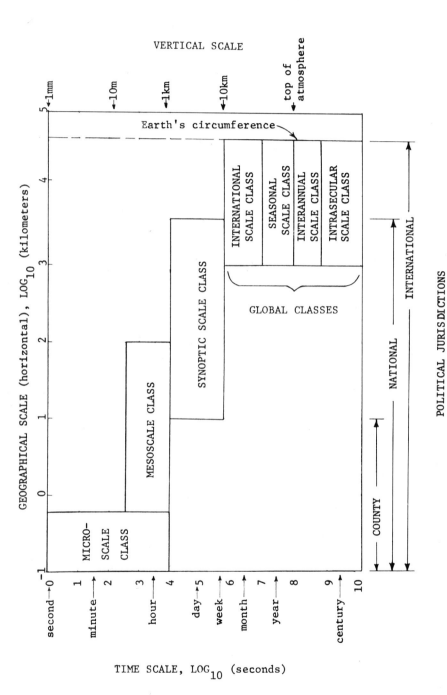

Figure 2. A scheme for scale classifications. The analyst first places his pesticide problem in one of these classifications. The classification is an aid to the selection of modeling methods.

$$\frac{\partial \bar{c}}{\partial t} + \bar{u}\frac{\partial \bar{c}}{\partial x} + \bar{v}\frac{\partial \bar{c}}{\partial y} + \bar{w}\frac{\partial \bar{c}}{\partial z} + \frac{1}{\rho}\left|\frac{\partial}{\partial x}\ \overline{\rho c'u'} + \frac{\partial}{\partial y}\ \overline{\rho v'c'} + \frac{\partial}{\partial z}\ \overline{\rho v'w'}\right| + S \quad (1)$$

where

   c = mixing ratio of substance (Kgrams substance per Kgram of air)

   u = velocity component in horizontal direction x

   v = velocity component in horizontal direction y

   w = velocity component in altitudinal direction z

   ρ = air density

   S = production and/or scavenging of the substance per unit mass of air.

The prime (') denotes fluctuation on a time scale less than the degree of resolution of time in the modeling effort. The over-bar (‾) denotes time averages over periods of time sufficiently long to obtain meaningful averages but short with respect to the time scale.

   The above equation is valid if:

   a.   The chemical potential of gaseous substances in air is not greatly different from its mixing ratio or, in the case of particulate substances, the size distribution of the particulate substance does not vary with time.

Other restrictions associated with the use of the continuity equation are the following:

   b.   The magnitude of scavenging ("sink") terms depends only upon c;

   c.   The covariances $\overline{c'u'}$, $\overline{c'v'}$, and $\overline{c'w'}$ can, in some way, be related to mean values of velocity.

In practice, some or all of the conditions will often be violated; the meteorological parameters ($\bar{u}$, u', etc.) can be hypothesized for a worst case analysis or can be measured directly as in an air monitoring program.

Atmospheric Transport Models Useful for Micro-Class Analysis

Estimating the exposure of farm workers to pesticides is an example of a micro-class problem for which a micro-class diffusion model would be appropriate.

Turner (1969) compiled a very useful workbook which illustrates the use of the so-called Gaussian plume models. He included 26 example calculations. A complete derivation of these models is given by Haltiner and Martin (1957). The required information is stability class, wind speed and standard deviations of angles of wind. Turner refers to Pasquill (1961) for more accurate models which require wind fluctuation measurements. Turner states that sampling times of about ten minutes are appropriate for the use of such models. This degree of resolution would be adequate for micro-class analysis. Ide (1971) makes important points on sampling time and scale resolution. The great advantage of this method is that the calculations are easily made without the aid of a computer. The great disadvantage of the method is that the formulas have only a very tenuous physical basis. Thus, the analysis must not be extrapolate beyond the condition for which empirical relations were derived.

There are circumstances for which it is known that Gaussian plume models do not apply. However, these conditions are usually associated with irregularities in terrain and meteorological conditions which are less likely to occur in the agricultural setting. See Williamson (1973) for an introductory and qualitative description of Gaussian plume models and their limitations.

Lamb (1971) has developed probabilistic models with a more compelling physical basis.

Air Pollution Models Useful for Meso-Class Pesticide Problems

An example of a meso-scale problem is the long-term impact of pesticides from Near Eastern agricultural development on the shrimp fisheries in the Persian Gulf. A "box" or "compartmental" model may prove to be a useful device for obtaining approximate estimates of exposure levels. With this technique, the region to be considered is thought of as a box bounded on the top by the inversion layer. It is assumed that vertical mixing rapidly spreads the pollutant to the inversion layer (Figure 3).

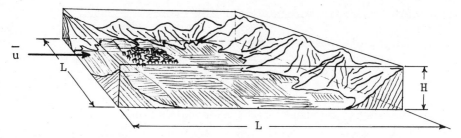

Figure 3. The single box model of atmospheric diffusion on the
meso-scale. Multi-box methods are discussed in the text.

In what follows, a derivation of the constraints on the resolution of time scales for box models will be followed by an example
application. Assume horizontal diffusion is negligible when compared to advection and, to within an order of magnitude,

$$\frac{\partial c}{\partial x} \simeq \frac{c}{L} \tag{2}$$

so that the continuity equation becomes

$$\frac{\partial c}{\partial t} + \frac{u}{L} c = S(t). \tag{3}$$

For the source term, take

$$S = \frac{Q}{H} - \gamma c; \qquad \gamma = \text{constant}, \tag{4}$$

where

$$Q = \text{total rate of production per unit of surface area.} \tag{5}$$

The second term on the right hand side of the equation is a scavenging term (e.g., disposition or first order chemical reaction with
an abundant co-reactant). The solution of the approximate continuity
equation becomes (Smith, 1961)

$$c = \frac{QL}{H(\overline{u} + L\gamma)} \left[1 - e^{\frac{(\overline{u}+L\gamma)t}{L}} \right]. \tag{6}$$

The usual recommendation (Smith, 1961; Sauter and Chilton, 1971; Crews, 1973) is that one approximate exposure level by

$$c \simeq \frac{QL}{Hu} \, . \tag{7}$$

When this recommendation is compared with the approximate solution of the continuity equations, the following conclusion is drawn:

> The minimum degree of resolution of the time scale must be greater than the time constant

$$T = \frac{L}{\overline{u} + L\gamma} \, . \tag{8}$$

The box model is illustrated with a study of the atmospheric lead aerosol in three California urban atmospheres conducted by one of the authors (Crews, 1973). First the matter of time scale. Sauter and Chilton estimate that the time constant of equation (8) is of the order of seven hours for the San Francisco Bay Area. This is a lower limit for the resolution of time scale. The characteristic dimension, L, is 49 kilometers for Los Angeles, 32.3 kilometers for San Diego, and 13 kilometers for Oakland.

Monthly average values of wind velocity and inversion height were obtained for the three areas from the Environmental Sciences Services Administration. The wind velocities are taken at six meters and are probably less than the true value of $\overline{u}$. The quantity Q was estimated from total yearly gasoline consumption figures (roughly eighty percent of the lead in gasoline becomes airborne) obtained for the state as a whole (Crews, 1973).

Equation (7) was then used to obtain predicted values of atmospheric lead concentration. These predictions (simulations) are compared with actual data in Figures 4 and 5. The comparison is surprisingly good.

Of course, if additional or more complete data is available, more accurate estimates can be obtained. Turner (1969) mentions the extension of plume models to longer time scales when "wind rose" (frequency diagrams for angular orientation) data is available. For a specific area, topographic irregularity may limit the usefulness of this method.

MacCracken et al. (1971) have demonstrated the accuracy of a multi-box model of carbon monoxide pollution in the San Francisco Bay Area. This type of model was developed for purposes of monitoring urban air pollution but the potential exists for monitoring airborne pesticides. Required input information is a fairly dense set of measurements of wind velocity and inversion heights in a

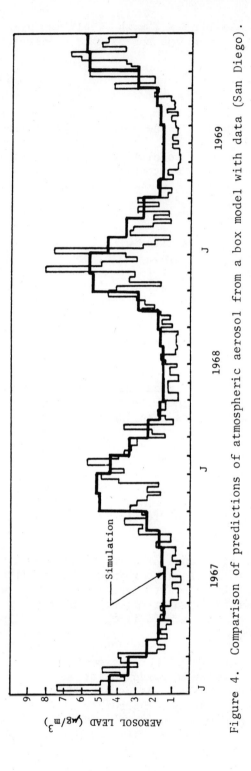

Figure 4.   Comparison of predictions of atmospheric aerosol from a box model with data (San Diego).

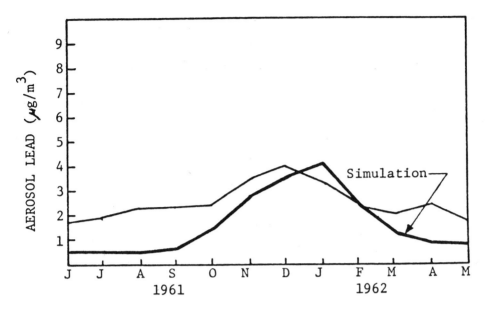

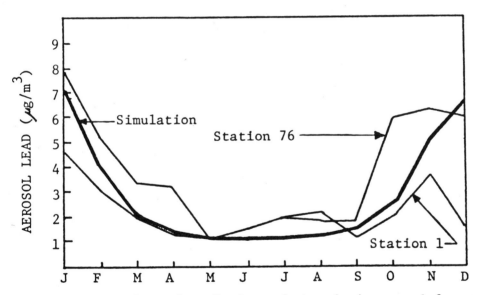

Figure 5. Comparison of predictions of atmospheric aerosol from a box model with data (Los Angeles).

meso-scale region. The technique employs a physical model to
generate a wind field and to track pollution transport. The significance of this work is that episodic (short time scale) estimates of exposure at the meso geographical scale can be obtained.
The Livermore group has indicated that they plan to extend the
model to pollutants which are not constrained by condition (4)
(i.e., to pollutants engaged in chemical reactions with other rare
constituents of the atmosphere). The use of these methods would
require the use of costly monitoring and computing equipment.

An equally important study has recently been reported upon by
Myrup and Morgan (1972). The goals and intent of this study seem
to be very similar to those of the Livermore study. Less meteorological input is required when using this method because the impact
of surface topology is modeled directly. The surface area of a
specific region is divided into land use categories which, in turn,
are associated with characteristic values of surface albedo, rough-
ness, heat capacity and other variables of meteorological significance. The land use categories are also associated with pollution
production rates and, thereby, the need for a data source inventory
is reduced. The adaptation of this method to the modeling of pesticide diffusion will offer few difficulties.

The last two techniques mentioned employ two alternative
methods for evaluating the meteorological parameters, $\bar{u}$, u', etc.
Another procedure is to solve the equations of motion of the atmosphere, thus generating the meteorological parameters. However,
in solving the equations of motion (three equations of momentum,
continuity, and energy), there is still a closure problem (more
unknowns than equations), because the atmosphere is turbulent.
Thus one must hypothesize a relationship between the mean flow and
the "Reynolds stresses" (u'v', w'v', u'w', etc.). A general procedure for solving the equations of motion for the atmosphere and
the diffusion equation for the pollutant concentration is to use a
finite difference technique. This method has been fairly well explained in Estoque and Bhumralkar (1970), but one should be cautioned
to consult the current literature since there are recently some con-
flicting results (Gutman, Torrance, and Estoque, 1973; Taylor and
DeLage, 1971). The basic procedure is to use an implicit finite
difference routine with an expanding grid system in altitude for
better differencing. A possible drawback with this type of approach
is that computer time may become extensive. The computer time is
a direct function of: 1) space scale of resolution; 2) time scale;
3) size of area being investigated; and 4) type of hypothesized
relationship in Reynold's stresses (Mellor and Herring, 1973).

Solution of (1) is the basis of other proposed methods of
transport analysis. Eschenroeder et al. (1972) demonstrate the
use of a model wherein a fixed column of air is followed in the
"Lagrangian sense." Vertical shear, transverse diffusion and

coriolis effects are neglected. The method provides quick estimates of pollutant concentration for the case of photochemical
reactions. The user is cautioned to study the basic assumptions.
Stevens et al. (1973) provide a method for studying photochemical
pollutant concentrations in a fixed grid or "Eulerian system." The
computer times for the latter method will be longer than for the
former but the assumptions are less restrictive and the "Eulerian"
output may be more useful.

Air Pollution Models Useful for Synoptic and International Class Problems

Starting with a discussion of the synoptic class, the review
paper by Munn and Bolin (1971) will be useful to any group interested in scale classification. This paper reviews, among other
important matters, constraints on scales and case histories of
pollution events. Munn and Bolin indicate the conditions under
which plume models will be useful at the synoptic scale. They indicate that this usefulness ends at the synoptic-international
interface. Their discussion indicates that the combination of
air trajectories with formulas for the spread of the pollutant,
derived from the theory of Fickian (i.e., constant eddy diffusivity)
diffusion could provide order of magnitude estimates. Munn and
Bolin are unwilling to come to this conclusion themselves.

Kwok et al. (1971) describe a simulation of carbon monoxide
transport at international and global scales using a finite difference representation of the continuity equation. Reiquam (1970)
extended the multi-box concept to a study of SO_2 transport at the
international scale. Reiquam concludes that high SO_2 concentration
in the Scandinavian atmosphere could indeed originate from sources
in Britain and Central Europe.

Air Pollution Models for Global Class Problems

Munn and Bolin (1971) indicate compelling reasons for assuming
that plume models are irrelevant at the global scale. One important reason is the significance and geographical distribution of
scavenging terms. They review physical chemical scavenging by
other abiotic components of the global environment. Robinson and
Robins (1968) review scavenging by chemical reaction within the
atmosphere itself.

The use of so-called reservoir models for global scale problems has been attempted by many authors. Keeling and Bolin (1967)
demonstrate the theoretical link of such models to the continuity
equation. They are interested in ocean models but their conclusion

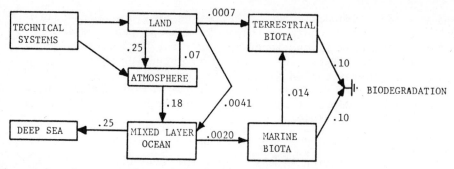

Figure 6. A reservoir model of the global transport of DDT. The yearly rate of change of DDT in each reservoir is taken to be proportional to the content of the donor reservoir. The proportionality constants are shown by the arrows. The biota reservoirs were added by Crews (1973) to the original model proposed by Woodwell et al. (1971).

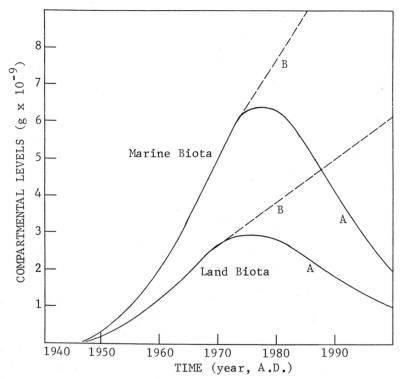

Figure 7. Simulation of the model indicated in Figure 6. A) Complete curtailment of DDT production; B) Continued increased production of DDT.

can be extrapolated to atmospheric models. Should reservoir models prove valid, an added advantage will be that reservoirs representing biotic and other abiotic elements can easily be added to the model. This fact has already been exploited by many authors (Crews, 1973; Eriksson and Welander, 1956; Young et al., 1971).

An example of a reservoir model of the global dynamics was developed by Woodwell et al. (1971) and extended by Crews (1973). The model is illustrated in Figure 6 and simulation results are displayed in Figure 7.

Other potentially more accurate methods, are at an even more speculative stage. Dwyer and Petersen (1973) have studied control volume procedures. Bolin and Keeling (1973) have made the first contribution to the study of the global continuity equation by eigen function expansion. Young (1974) has proposed important extensions of the method of Bolin and Keeling.

SUMMARY

In order to analyze the environmental dynamics of pesticides, it will be necessary to classify potential problems. The classification will depend upon level of toxicity, persistence, volatility, and other physical-chemical-biological properties. The authors have proposed that the classification be made on the basis of time and geographical scales of the potential problem and that the classification nomenclature coincide with the classification of the modeling technique to be used to estimate exposure levels. The classification scheme was illustrated, in a highly tentative way, in a literature survey of air pollution models by the authors.

REFERENCES

Bolin, B., and C.D. Keeling. 1963. Large scale atmospheric
    mixing as deduced from the seasonal and meridional variations
    of carbon dioxide. J. Geophys. Res. 68(13).

Crews, W.B. 1973. Preliminary static and dynamic models for the
    global transport of lead and DDT. M.S. Thesis, Mechanical
    Engineering, University of California, Davis.

Dwyer, H.A., and T.J. Petersen. 1973. Time dependent global
    energy modeling. J. Appl. Meteorol. Feb. 36-42.

Eriksson, E., and P. Welander. 1956. On a mathematical model of
    the carbon cycle in nature. Tellus VIII(2):155-175.

Eschenroeder, A.Q., Martinez, and R.A. Nordsieck. 1972. Evalua-
    tion of a diffusion model for photochemical smog simulation.
    General Research Corporation, Santa Barbara, California, GP-1-
    273, October.

Estoque, M.A., and C.M. Bhumralkar. 1970. A method for solving
    the planetary boundary layer equations. Boundary Layer
    Meteorology 1:169-194.

Gutman, P.P., K.E. Torrance, and M.A. Estoque. 1973. Use of the
    numerical method of Estoque and Bhumralkar for the planetary
    boundary layer. Boundary Layer Meteorology 5:341-346.

Haltiner, G.J., and F.L. Martin. 1957. Dynamical and physical
    meteorology. McGraw-Hill, Chapter 16.

Ide, Yasuo. 1971. The effect of observation time on turbulent
    diffusion in the atmosphere. Water, Air, and Soil Pollution
    1L32-41.

Keeling, C.D., and B. Bolin. 1967. The simultaneous use of
    chemical tracers in oceanic studies: I, general theory of
    reservoir models. Tellus XIX(4):566-581.

Kwok, H.C.W., W.E. Langlois, and R.A. Ellefsen. 1971. Digital
    simulation of the global transport of carbon monoxide. J.
    Res. Devel. (IBM) 15(1).

Lamb, R.G. 1971. Numerical modeling of urban air pollution.
    Ph.D. Thesis, Department of Meteorology, University of
    California, Los Angeles.

MacCracken, M.C., T.V. Crawford, K.R. Peterson, and J.B. Knox.
    1971. Development of a multi-box air pollution model and
    initial validation for the San Francisco Bay Area. URCL-
    73348, Lawrence Radiation Laboratory, Livermore, California.

Mellor, G.L., and H.J. Herring. 1973. A survey of the mean tur-
    bulent field closure models. AIAA Journal 11(5).

Monin, A.S. 1972. Weather forecasting as a problem in physics.
    MIT Press, Chapter 1.

Munn, R.E., and B. Bolin. 1971. Global air pollution--meteoro-
    logical aspects. Atmospheric Environment, Pergamon Press
    5:363-402.

Myrup, L.O., and D.L. Morgan. 1972. Numerical model of the urban
    atmosphere, Vol. 1. The City Interface, Contribution in At-
    mospheric Science, No. 4, Department of Agricultural Engin-
    eering, University of California, Davis.

Pasquill, F. 1961. The estimation of the dispersion of wind
    borne material. Meteorology Magazine 90(1063):33-49.

Reiquam, H. 1970. Sulfur, simulated long-range transport in the
    atmosphere. Science 170:318-320.

Robinson, E., and R.C. Robins. 1968. Sources, abundance, and fate
    of gaseous atmospheric pollutants. SRI Project PR-6755,
    Report N71-25147, Stanford Research Institute.

Sauter, G.D., and E.G. Chilton. 1970. Air improvement recommenda-
    tions for the San Francisco Bay Area. Stanford-Ames NASA-ASEE
    Summer Faculty Workshop.

Smith, M.E. 1961. The concentrations and residence times of
    pollutants in the atmosphere. Chapter 10 in Chemical Re-
    searchers in the Lower Atmosphere, Interscience.

Stevens, S.D., M. Liu, T.A. Hecht, P.M. Roth, and J.H. Seinfeld.
    1973. Further development and evaluation of a simulation
    model for estimating ground level concentrations of photo-
    chemical pollutants. Systems Applications Inc., Beverly
    Hills, California, R-73-19, February.

Sutton, O.G. 1932. A theory of eddy diffusion in the atmosphere.
    Proceedings of the Royal Society. Series A, 135:143-165.

Taylor, P.A., and Y. DeLage. 1971. A note on finite-difference
    schemes for the surface and boundary layers. Boundary layer
    meteorology, 108-121.

Turner, D.B.   1969.   Workbook of atmospheric dispersion estimates.
    U.S. Department of Health, Education, and Welfare.   National
    Air Pollution Control Administration, Cincinnati, Ohio.

Williamson, S.J.   1973.   Fundamentals of air pollution, Addison-
    Wesley, Chapter 7.

Woodwell, G.M., P.P. Craig, and H.A. Johnson.   1971.   DDT in the
    biosphere: where does it go? Science 174.

Young, J.W., J.W. Brewer, and J.M. Jameel.   1972.   An engineering
    systems analysis of man's impact on the global carbon cycle.
    Systems and Simulation in the Service of Society.   Simulation
    Council, 1,2:31-40.

Young, J.W.   1974.   Model simplification of a system of bilaterally
    coupled diffusive elements: with applications to global
    atmospheric pollutant transport problems.   Ph.D. Dissertation,
    Department of Mechanical Engineering, University of California,
    Davis, California.

ROLE OF ADSORPTION IN STUDYING THE DYNAMICS OF PESTICIDES IN A SOIL ENVIRONMENT

R. Haque
Department of Agricultural Chemistry and Environmental
Health Sciences Center, Oregon State University
Corvallis, Oregon 97331

Once a pesticide or a toxic chemical finds its way in the environment a major part of it comes in contact with soils. In many instances soil acts as a sink for many chemicals. Thus in order to understand the behavior of pesticides in the environment we must know their dynamics in the soil. The three important factors controlling the behavior of a pesticide in a soil environment are: 1) the sorption/desorption process; 2) leaching-diffusion; and 3) degradation. The subject of leaching, diffusion and degradation will be discussed in the next two chapters. In this manuscript we shall present some of the important points in studying the adsorption of pesticides to soil colloids. We shall also describe how the adsorption process may influence the dynamics of pesticide in a soil environment.

A typical soil environment represents solid, liquid as well as gaseous phases (Marshall, 1964; Bear, 1965). The solid includes minerals and organic matters whereas the water and dissolved salts constitute the liquid phase and gases such as nitrogen, oxygen, carbon dioxide and trace rare gases are the part of gaseous phase. Thus a chemical in soil will represent three distinct transport problems. For the purpose of adsorption we should know more about solid phase. The main constituent representing soils are clay material, organic matter, oxides and hydroxides of aluminum and silicon. A discussion of the nature of the solid constituents of the soil such as clays, organic matter and oxides will be of importance to the adsorption process.

Clays represent layers of silica and aluminum sheets (Van Olphen, 1963; Grim, 1968). The silica consists of a silicon atom surrounded by four oxygen atoms in a tetrahedral symmetry. Alumina represents aluminum atoms coordinated by six oxygen or hydroxyl groups in an octahedral fashion. In many cases the aluminum and silicon atoms are replaced by such atoms as iron, manganese and magnesium. These substitutions produce a change in the net charge on the clay surface. In the clay material the silica and alumina sheet are usually in a 1:1 or 2:1 ratio. A clay whose one silica and one alumina unit form a layer sheet is best represented by kaolinite. Other clays such as montmorillonite, illite and vermiculite are examples of 2:1 silica-alumina sheet. Clays are characterized by such properties as lattice expansion, cation exchange capacity, surface area, etc. For example montmorillonite is an expanding clay it swells appreciably in presence of water whereas kaolinite does not and is a non-expanding clay. Some of the characteristics of the three important clays; montmorillonite, kaolinite and illite are given in Table 1. Water plays an important role in defining the characteristics of a clay (Low, 1961). Adsorbed water on the clay surface is usually more ordered than in the bulk solution. It has also been suggested that water on the clay surface is more ionized than in bulk solution, thus the acidity on the surface is higher than in solution (Mortland et al., 1963).

Organic matter of the soils commonly known as humic substances is another important constituent of soil and are formed by the degradation of plant and animal tissues. The major constituents of humic acids are polymeric organic acids which are free radicals. The minor constituents of humic acid are legnin, proteins, enzymes, sugars, fat amino acid, resin, wax, etc. Because of the complex nature of the chemical structure of humic acids, these are characterized by their solubility in an alkali or acid. An acid component is commonly known as fulvic acid whereas humic acid is base soluble. Humin is the name given to the insoluble constituent. A typical structure of humic acid proposed by Kononova (1966) is given as follows:

HUMIC ACID

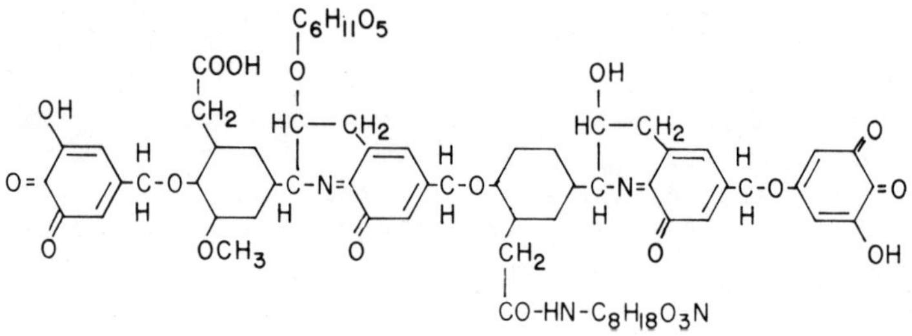

TABLE 1

Characteristics of Three Clays

Clay	Type	Surface Area m^2/g	Cation Exchange Capacity, meq/100g
Kaolinite (non-expanding)	1:1	25-50	2-10
Montmorillonite (expanding)	2:1	700-750	80-120
Illite	2:1	75-125	15-40

Humic substances usually possess large surface area (500-800 m^2/g) and cation exchange capacity in the range of 200-400 meg/100g. The large surface area and the presence of various function groups is probably responsible for the high adsorption capacity of the soil organic matter.

Amorphous mineral colloids such as oxides of aluminum, silicon and iron are also important in considering the adsorption of organic compounds on soils. A material substance comprising of amorphous-silica and alumina mixture with iron substitution is also of importance in this regard.

A chemical such as a pesticide molecule will be adsorbed to soil as soon as it comes in contact with the surface. The adsorption will depend upon the energy of the surfaces representing various constituents of the soil. Factors affecting the adsorption of pesticides on soils has been discussed recently by various workers (Bailey and White, 1964; Hamaker and Thompson, 1972; Freed and Haque, 1973). Although the adsorption experiments on soil-pesticide adsorption are relatively simple to perform, the results are usually difficult to interpret. The adsorption behavior (Osipow, 1962) of a pesticide on a soil surface can be visualized with the aid of Leonard-Jones (1932) potential energy diagram (Figure 1). As the distance of the adsorbate to the surface is decreased there is attraction, with a further decrease the potential energy increases. Here HA is the heat of adsorption. Line I is for a physical or Vander-Waal type adsorption whereas II is for chemisorption. At a point where both the graphs intersect is the activation energy. Usually the chemisorption process require

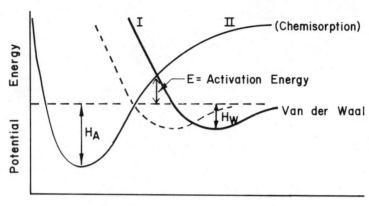

Distance from Surface to Absorbed Unit

Figure 1. Adsorption process as represented by a potential energy diagram (Leonard-Jones, 1932; Osipow, 1962).

higher activation energy, and the heat of adsorption for chemisorption HA is always greater than the heat of adsorption for physisorption. For a physisorption the activation energy as well as the heat of adsorption are quite low. A pesticide-soil system generally follows a physical type adsorption.

The adsorption data are generally represented by an equation known as isotherm. An isotherm represents a relation between the amount of chemical sorbed and the chemical left in the solution after equilibrium. The three well-known isotherms are Freundlich, Langmiur and BET (Brunauer, Emmett and Teller). The Freundlich type isotherm is based on an empirical equation, Langmiur on a monolayer formation and BET on multilayer adsorption considerations. These isotherms are given as follows:

Freundlich: $\dfrac{x}{m} = KC^{n}$ (1)

Langmiur: $\dfrac{x}{m} = \dfrac{abC}{1+aC}$ (2)

BET: $\dfrac{C}{(C_{o}-C)X} = \dfrac{Km^{-1}}{bKm} \cdot \dfrac{C}{C_{o}} = \dfrac{1}{bKm}$ (3)

In the above equations K and n are constants, C_{o} = maximum amount of chemical available for adsorption, C = equilibrium concentration; x = amount of chemical sorbed; m = amount of surface material or soil; a and b are constants; and Km is defined by the following equation:

$$Km = \frac{Q_1 - Q_L}{RT} \qquad (4)$$

where Q_1 and Q_L are molar heat of adsorption and crystallization, R the gas constant and T the absolute temperature. Majority of data representing the adsorption of pesticides on soils are represented by Freundlich type isotherm. Langmiur and BET isotherms are very little used for soil-pesticide system. In practice the logarithm of chemical adsorbed is plotted against the logarithm of chemical at equilibrium. Such a plot usually results in a straight line and its slope and intercept give the value of constants n and K. The constant K represents the extent of adsorption. Whereas n throws much light on the nature of the adsorption as well as the role of solvent (water) is the adsorption. The strength of adsorption is determined by the temperature dependence of adsorption data.

The kinetics of adsorption is also an important factor. For most adsorption process involving soil-pesticides the adsorption process is fast. Usually half an hour is required for the completion of most part of the adsorption. Although longer times are needed to complete the final portion of the adsorption. The kinetics of pesticide adsorption has been described by various workers using different approaches (Morris and Weber, 1966; Haque et al., 1968; Lindstrom et al., 1970)

Since pesticides include a large number of compounds, it will be impossible to describe the adsorption characteristics of each in this manuscript. However, we shall discuss the adsorption characteristics of three classes of pesticides and related compounds (Table 2). These chemicals are selected because of their structural characteristics, properties and environmental significance. The three classes include the quaternary bipyridimium salts, chlorinated phenoxy acetic acid and chlorinated aromatic hydrocarbons.

Diquat and paraquat are the two important members of the bipyridimium quaternary salts. These chemicals possess high water solubility (Brian et al., 1958) and are readily adsorbed on clay and soil surfaces. Weber (1968) has described the adsorption of these chemicals on clay surfaces. It is generally believed that the adsorption of diquat or paraquat from aqueous solution on a clay or soil surface involves the cation exchange reaction.

$$Diquat^{2+} + 2Na\text{-clay} \rightarrow Diquat\text{-clay} + 2Na^{+} \qquad (5)$$

TABLE 2

Selected Chemicals

Chemical	Formula	Solubility Characteristics
Quaternary Bipyridimium Salts		
Diquat	2 Br⁻	High water solubility
Paraquat	2 Cl⁻	High water solubility
Phenoxy Acetic Acid		
2,4-D		Moderate solubility (500-600 ppm)
Chlorinated Hydrocarbons		
DDT		Extremely low water solubility
Polychlorinated Biphenyl		Low solubility - depending on the number of chlorine atoms

The preceding reaction (5) is strongly to the right side. However, the diquat or paraquat cations may be removed from the surface. The ease of removal depends upon the surface. For the expanding clay such as montmorillonite the displacement of diquat and paraquat from the clay is rather difficult. X-ray diffraction studies have shown that these cations penetrate into the interlamellar spacing layer of montmorillonite clay (Weed and Weber, 1968; Weber et al., 1965). Recent studies of Haque et al. (1970) have shown that diquat and paraquat adsorption on montmorillonite surface involves a charge-transfer type mechanism. This conclusion was reached on the basis of ultra-violet and infra-red spectroscopic investigations of the diquat and paraquat on montmorillonite surface. It is well established that diquat and paraquat form charge-transfer complexes with halide ions (Haque et al., 1969; Haque and Lilley, 1972). The addition of montmorillonite to diquat or paraquat solution showed distinct changes in the adsorption maxima of the herbicide. Both diquat and paraquat showed a red shift when the montmorillonite was added indicating an interaction between the herbicide and the clay. The ultra-violet changes are given in Table 3. The infra-red spectra of diquat and paraquat in a clay matrix also showed the evidence of the formation of complexes. The infra-red spectra of diquat and paraquat especially the out-of-place C-H vibration frequencies in the region 600-800 cm^{-1} showed changes when it was adsorbed to the clay surface (Figure 2). The strongest

TABLE 3

Ultra Violet Absorption Maxima
of Diquat and Paraquat (5 X 10^{-5} M)
in the Presence of Montmorillonite[a]

Adsorbent	Concentration of Montmorillonite mg/liter	Maximum Wavelength
Diquat	0	309, 318
	200	317, 327
Paraquat	0	258
	200	278

[a]Haque et al. (1970).

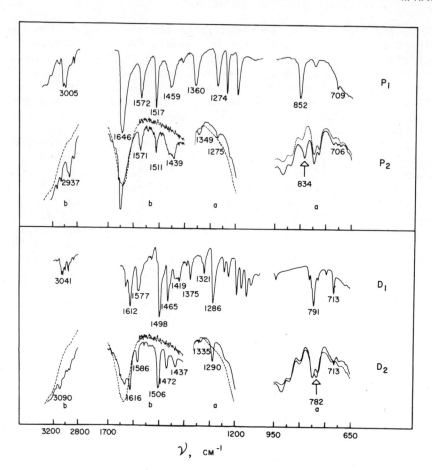

Figure 2. Infra-red spectra of diquat and paraquat on montmorillonite surface (Haque et al., 1970).

band in this region, the out-of-place C-H commonly known as umbrella, mode shifted to 834 cm^{-1} from 854 cm^{-1} for paraquat clay complex and 782 cm^{-1} to 793 cm^{-1} for diquat complex. It has been shown that chloride, bromide and carbide ions show similar shifts (Haque and Lilley, 1972) and is attributed to a charge-transfer type complex formation between the anion and the diquat or paraquat cation. It may be possible that montmorillonite may be acting as strong enough electron donating so that diquat or paraquat may be forming charge-transfer complexes. These studies suggest that in pesticide-soil interaction charge transfer complex formation may be possible. This is the first demonstration of charge-transfer complex formation in pesticide-soil system.

The other class of compound mentioned in Table 2 is phenoxy type. 2,4-D and 2,4,5-T are the well known examples. These chemicals are important herbicides and possess water solubility in the range of few hundred parts per million. These chemicals are also adsorbed on clay and soil surfaces (Weber, 1972) although the extent of adsorption is much less as compared to diquat and paraquat. The adsorption of 2,4-D on material surfaces can be represented by the Freundlich type isotherm (Haque and Sexton, 1968). The Freundlich constants for the adsorption of 2,4-D on mineral surfaces is given in Table 4. It is interesting to note that the constant K which is an indirect measure of the extent of adsorption is strongly dependent upon the nature of the surface. The value of K is lowest for sand and highest for humic acid. The other surface has a K value in between. The high value of K is due to the fact that humic acid being organic has larger surface area and reaction sites on the surface. Sand possessing very few active sites is a very poor sorber. These findings are consistent with earlier theories that organic matter of a soil is important

TABLE 4

Freundlich Isotherm Constant K and n
for the Adsorption of 2,4-D on Surfaces[a]

Surface	Temperature °C	n	log K
Illite	0	0.685	1.11
	25	0.719	1.02
Montmorillonite	0	.925	-0.063
	25	1.004	-0.186
Sand	0	.671	- .984
	25	.827	-1.454
Alumina	0	.97	-0.06
	25	1.01	-0.08
Silica gel	0	.90	0.58
	25	.95	.11
Humic acid	0	0.86	2.01
	25	0.931	1.9

[a]Haque and Sexton (1968).

in describing absorption (Lambert, 1968; Haque and Coshow, 1971).
The K value increases with decreasing temperature indicating that
the adsorption decreases with increasing temperature. This is
supporting the fact that the adsorption is an exothermic process.

The heat of adsorption ΔH was calculated as a function of
amount of chemical sorbed using the following equation:

$$(\Delta H)_x = R \left[\frac{\delta \ln C}{\delta (1/T)}\right] \tag{6}$$

showed that for most surfaces the ΔH value became more positive
with increasing temperature. For most surfaces the heat of
adsorption was very small 1-2 K · cal/mole indicating string
binding. This indicated that the adsorption of 2,4-D on most
surfaces is a physical type. Only in the case of sand and silica
gel is it evident that the binding is much stronger. A hydrogen
bond formation between 2,4-D and silica group on sand and silica
gel as shown below is suggested.

$$-Si \underset{OH- - - -O}{\overset{O- - - -HO}{\diagdown\diagup}} C-CH_2-O-C_6H_4Cl_2$$

The last class of compounds known as chlorinated hydrocarbons
are characterized by their very strong hydrophobic character. The
insecticide DDT and the industrial chemicals polychlorinated biphenyls PCB are the important members in this group. These chemicals
possess very low water solubility (in the parts per billion range).
Because of this property the chlorinated hydrocarbons are readily
adsorbed on particulate matter and soils. This characteristic
plays an important mechanism in the transport of such chemicals
in the biosphere. The low solubility of chlorinated hydrocarbons
in water presents many difficulties in carrying out conventional
adsorption studies. Shin et al. (1970) have described the adsorption
of DDT on a variety of surfaces. However, they used suspension of
DDT of much higher concentration rather than an equilibrium solution
possessing DDT in its solubility range. Guenzi and Beard (1967)
have discussed the adsorption of DDT in terms of its precipitation
on surfaces. However, in order to describe the adsorption of
chlorinated hydrocarbons which may have significance one must use
the adsorbate solution equal to or less than its solubility limit.
Haque et al. (1974) have recently described the adsorption of the
polychlorinated biphenyls Aroclor[R] 1254 and 2,4,2',4' tetrachloro-
biphenyl (Haque and Schmedding, 1974) on different surfaces from
aqueous solution. The PCB solution used was a saturated one in
water. The addition of various amounts of adsorbant in a saturated
aqueous soluton of Aroclor[R] 1254 decreased its concentration. The
decrease in concentration was attributed to adsorption (Figure 3).

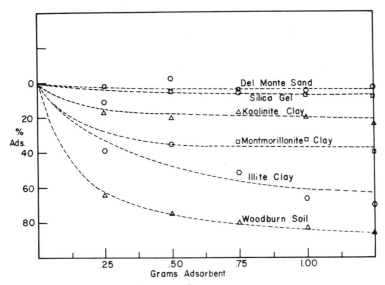

Figure 3. Adsorption of Aroclor[R] 1254 on various surfaces (Haque
et al., 1974).

It is interesting to note that Woodburn soil is the best adsorber
whereas sand is a poor adsorber. This is consistent with the
earlier finding where the organic matter in the soil caused more
adsorption as compared to the fewer sites on sand surface. The
Freundlich type plot for the adsorption of Aroclor[R] 1254 on Woodburn soil and illite clay surface is shown in Figure 4. The value
of K and n for Woodburn soil and illite clay are 63.1 and 1.1 and
26.3 and 0.81 respectively. It is apparent that K for Woodburn
soil is higher than illite.

The adsorption of 2,4,2',4' tetrachlorbiphenyl on clay,
soil, sand and humic acid surfaces shows similar behavior (Haque and
Schmedding, 1974). The adsorption was highest on humic acid surface
and was very slow on sand. The Freundlich type isotherm is shown
in Figure 5. Again the constant K is highest for humic acid
indicating the strong dependence of adsorption on organic matter.
Desorption studies on these surfaces indicated that 2,4,2',4'
tetrachlorbiphenyl did not desorb from a surface very easily. Very
insignificant amounts of chemical were desorbed, indicating a
tight binding and strong hydrophobic nature of the chemicals.

From the above adsorption studies it is apparent that the
adsorption of pesticides and related chemical on surfaces is

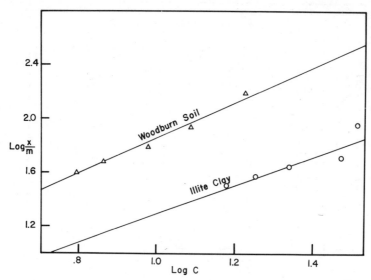

Figure 4. Freundlich plot for the adsorption of Aroclor[R] 1254 on illite and Woodburn soil surface (Haque et al., 1974).

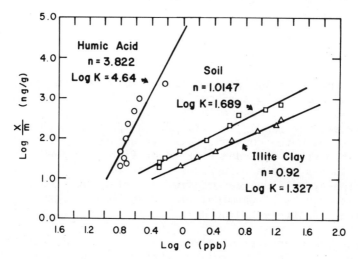

Figure 5. Freundlich plot for the adsorption of 2,4,2',4' tetra-dichlorobiphenyl on illite, humic acid and Woodburn soil surface (Haque and Schmedding, 1974).

dependent on many factors such as solubility of the pesticide, chemical structure, nature of the surface, organic matter, etc.

ROLE OF ADSORPTION IN PREDICTING THE DYNAMICS OF PESTICIDES IN SOILS

The three important factors involved in the dynamics of pesticides in soils are: 1) leaching; 2) diffusion; and 3) vapor loss. The leaching and diffusion will describe the movement in soil whereas vapor loss can tell how the chemical will transport in the air from the soil. Adsorption will undoubtedly effect very strongly all the above three processes. The effect of adsorption on the leaching has recently been demonstrated by Saxena (1972). In this experiment increasing amount of resin was added to glass beads and the leaching of 2,4-D was monitored as a function of resin concentration. It was found that the leaching of 2,4-D was retarded as the resin concentration was increased. Since resin provided the adsorption site this is a clear demonstration that increasing adsorption retards the leaching. The diffusion of pesticides in soils is also influenced very strongly with the adsorption. Lindstrom et al. (1969) have measured the diffusion coefficient of 2,4-D in a variety of soils and have found a correlation between soil organic matter and the diffusion coefficient. The diffusion of pesticides in soils was strongly dependent upon the organic content of soil. The diffusion coefficient decreased with increasing organic content of the soil.

The vapor loss of pesticides and related compounds also depends strongly upon the adsorption to soils. Recent studies of Haque et al. (1974) have shown that the vapor loss of Aroclor[R] 1254 from a sand surface was significant (Figure 6). This could be due to poorly adsorbing capacity of sand. However, when similar experiments were carried out for the vapor loss of the Aroclor[R] 1254 adsorbed on a soil surface the loss was insignificant (Figure 7). This demonstrated that adsorption of a chemical on surfaces retarded the vapor loss.

SUMMARY

Adsorption process plays an important role in determining the fate and behavior of pesticides and related chemicals in a soil environment. Major factors influencing the adsorption-desorption include water solubility of the chemical, nature of the soil constituents, pH, temperature, molecular structure and hydrophobicity of the pesticide. The adsorption of highly soluble pesticides such as organic cation diquat and paraquat on clay surfaces involves an exchange mechanism. Spectroscopic investigations have shown

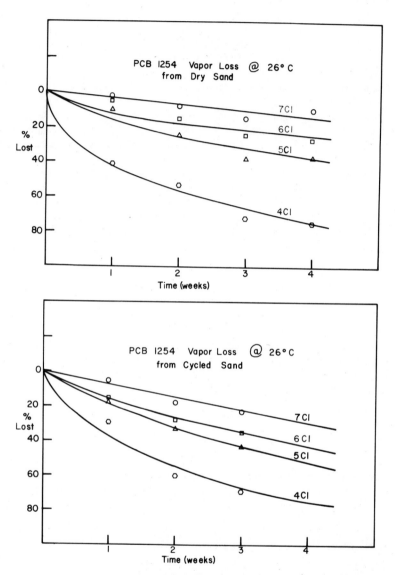

Figure 6.   Vapor loss of Aroclor[R] 1254 from sand surfaces (Haque et al., 1974).

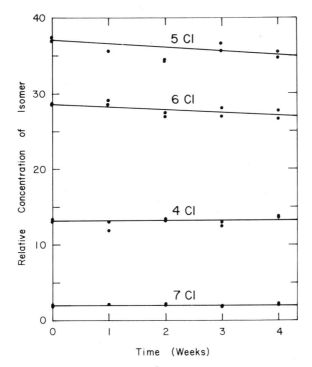

Figure 7.   Vapor loss of Aroclor[R] 1254 from Woodburn soil surface.

that a charge-transfer type complex is formed between montmorillonite
and diquat or paraquat. The moderately water soluble pesticide
2,4-D is physically adsorbed on most of the surfaces constituting
a soil. The adsorption data can best be represented by Freundlich
type isotherm. The hydrophobic chlorinated hydrocarbon compounds
such as DDT and PCB's are strongly adsorbed to surfaces. A method
for studying the adsorption of hydrophobic compounds such as DDT
and PCB's on surface material is described.

The adsorption greatly influences the dynamics of pesticides
in soils. The diffusion, leaching and vapor loss of pesticides in
soils is retarded as the extent of adsorption increases. The extent
of adsorption increases directly with an increase in the organic
content of the soil. Thus, an increase in organic content of a
soil may greatly reduce the diffusion, leaching and vapor loss of
pesticides in soils.

ACKNOWLEDGEMENTS

Some of the work described in this manuscript has been supported by grant ES-00040 fron the National Institute of Environmental Health Sciences. The author wishes to thank W.R. Coshow, S. Lilley, D. Schmedding and R. Sexton for their excellent technical assistance over the years in connection with some of the work reported in this paper.

REFERENCES

Bailey, G.W. and J.L. White. 1964. Review of adsorption and desorption of organic pesticides by soil colloids, with implications concerning pesticide bioactivity. J. Ag. Food Chem. 12:324.

Bear, F.E. 1965. Chemistry of the Soil, Reinhold Publishing Corporation, N.Y.

Brian, R.C., R.F. Homer, J. Stubbs and R.L. Jones. 1958. A new herbicide I: 1'-ethylene-2,2'-dipyridylium dibromide. Nature 181:446.

Freed, V.H. and R. Haque. 1973. Adsorption, movement and distribution of pesticides in soils. In Pesticide Formulations. W. Van Valkenburg, editor. Marcell-Dekker, N.Y. p. 441.

Grim, R.E. 1968. Clay Minerology, McGraw Hill, N.Y.

Guenzi, W.D. and N.E. Beard. 1967. Movement and persistence of DDT and lindane in soil columns. Soil Sci. Soc. Amer Proc. 31:644.

Hamaker, J.W. and J.M. Thompson. 1972. Adsorption. In Organic Chemicals in the Soil Environment, Volume I. C.A.I. Goring and J.W. Hamaker, editors. Marcell-Dekker, N.Y.

Haque, R. and W.R. Coshow. 1971. Adsorption of isocil and bromacil from aqueous solution onto some mineral surfaces. Env. Sci. and Technol. 5:139.

Haque, R., W.R. Coshow, and L.F. Johnson 1969. NMR studies of diquat, paraquat and their charge transfer complexes. J. Am. Chem. Soc. 91:3822.

Haque, R. and S. Lilley. 1972. Infrared spectroscopic studies of charge-transfer complexes of diquat and paraquat. J. Ag. Food Chem. 20:57.

Haque, R., S. Lilley and W.R. Coshow. 1970. Mechanism of adsorption of diquat and paraquat on montmorillonite surface. J. Coll. Int. Sci. 33:185.

Haque, R., F.T. Lindstrom, V.H. Freed and R. Sexton. 1968. Kinetic study of the sorption of 2,4-D on some clays. Env. Sci. and Technol. 2:207.

Haque, R. and D. Schmedding. 1974. Desorption of some PCB isomers studies on the water solubility and adsorption. Proc. 29th N.W. Regional Meeting. American Chemical Society. p. 15.

Haque, R., D. Schmedding and V.H. Freed. 1974. Aqueous solubility, adsorption and vapor behavior of polychlorinated biphenyl Aroclor 1254. Environ. Sci. and Technol. 8:139.

Haque, R. and R. Sexton. 1968. Kinetic and equilibrium study of the adsorption of 2,4-D dichlorophenoxy acetic acid on some surfaces. J. Coll. Int. Sci. 27:818.

Kononova, M.M. 1966. Soil Organic Matter. Pergamon Press, N.Y.

Lambert, S.M. 1968. Omega, a useful index of soil sorption equilibria. J. Ag. Food Chem. 16:340.

Leonard-Jones, J.E. 1932. Processes of adsorption and diffusion on solid surfaces. Trans. Faraday Soc. 28:334.

Lindstrom, F.T., L. Boersma and H. Gardiner. 1968. 2,4-D difussion in saturated soils: A mathematical theory. Soil Sci. 106:107.

Lindstrom, F.T., R. Haque and W.R. Coshow. 1970. Adsorption from solution III a new model for the kinetics of adsorption-desorption process. J. Phys. Chem. 74:495.

Low, P.F. 1961. Physical chemistry of clay-water interaction. Adv. Agron. 13:269.

Marshall, C.E. 1964. The Physical Chemistry and Minerology of Soils. John Wiley and Sons, Inc., N.Y.

Morris, J.C. and W.J. Weber. 1966. Adsorption of biochemically resistant materials from solution 2. Env. Health. Sci. AWTR 16:133.

Mortland, M. 1963. Interaction between ammonia and the expanding lattices of montmorillonite and vermiculite. J. Phys. Chem. 67:248.

Osipow, L.I. Surface Chemistry in Theory and Industrial
    Applications. Reinhold Publishing Corporation, N.Y.

Saxena, S.K. 1972. Theorietical and experimental evaluation of
    transfer 2,4-dichlorophenoxyacetic acid in porous media. Ph.D.
    Thesis, Oregon State University, Corvallis, Oregon.

Shin, Y., J.J. Chodan and A.R. Wolcott. 1970. Sorption of DDT
    by soils, soil fractions and biological materials. J. Ag.
    Food Chem. 18:1129.

Van Olphen, J. 1963. An Introduction to Clay Colloid Chemistry
    for Clay Technologists, Geologists and Soil Scientists.
    New York Interscience Publications.

Weber, J.B. 1972. Interaction of organic pesticides with
    particulate matter in aquatic soil system. Adv. in
    Chem. Ser. 55.

Weber, J.B. and H.D. Coble. 1968. Microbial decomposition of
    diquat adsorbed on montmorillonite and kaolinite clays.
    J. Ag. Food Chem. 16:475.

Weber, J.B., P.W. Perry and R.P. Upchurch. 1965. The influence
    on temperature and time on the adsorption of paraquat,
    diquat, 2,4-D and prometone by charcoal and anion exchange
    resin. Soil Sci. Soc. Amer. Proc. 29:678

Weber, J.B. and D.C. Scott. 1966. Availability of a cationic
    herbicide adsorbed on clay minerals to cucumber seedlings.
    Science 152:1400.

Weber, J.B. and S.B. Weed. 1968. Adsorption and desorption of
    diquat, paraquat and prometone by montmorillonite and
    kaolinite clay minerals. Soil Sci. Soc. Amer. Proc.
    32:485.

Weed, S.B. and J.B. Weber. 1968. The effect of adsorbent charge
    on the competitive adsorption of divalent organic cations
    by layer-silicate minerals. Amer. Minerologist 53:478.

THE INTERPRETATION OF SOIL LEACHING EXPERIMENTS

John W. Hamaker
Ag-Organics Research Department
Dow Chemical U.S.A.
Walnut Creek, California 94598

INTRODUCTION

Leaching of chemicals through soil is an environmental .concern because of the possibility that they will reach the water table and contaminate the ground water. However, whether a chemical will reach the ground water will depend not only upon its movement through the soil, but also upon its disappearance from the soil. If, for instance, the rate of degradation is sufficiently rapid compared to the rate of leaching, the chemical will disappear before it can reach the ground water and, therefore, will not pose that environmental problem. The determination of soil leaching rates is important because the rate of leaching of a chemical indicates how long a chemical is retained in the top soil where it is most subject to degradation or dissipation. It is important to consider the rate of degradation, but, in spite of this, there have been very few efforts to deal with the problem of simultaneous degradation and leaching (King and McCarty, 1968; Lehav and Kahanovitch). The environmental significance of soil leaching of pesticides will not be properly understood until this is done. This paper discusses only the leaching aspect of the problem.

As it occurs in the real world, leaching through soil is a complex phenomenon with at least four major elements:

1. Soil adsorption which determines the underlying pattern;

2. Porous flow and diffusion which produce a dispersion of the chemical;

3. Adsorption dynamics which introduces a factor of hysteresis; and

4. Water infiltration and evaporation which determine the actual amount of water movement and, hence, the observed amount of movement of a chemical.

Soil adsorption is the property of the chemical which is responsible for the basic pattern of movement of chemicals through soil. The time spent by the chemical in the adsorbed condition causes it to lag behind the solvent. If the adsorption equilibrium is rapid enough, the initial pattern will be maintained, as for example, a spot or zone, i.e., chromatographic movement.

This chromatographic pattern, however, is modified by the phenomenon of flow through a porous medium which acts to both disperse and retard the chemical. This has been termed "hydrodynamic dispersion" because, unlike diffusion which also disperses the chemical, it is a dynamic effect depending upon movement of the liquid througn pores. Some of the liquid moves a greater distance through more tortuous passages or more slowly through smaller passages and accomplishes less direct line movement than a solution taking a more direct path. In addition, chemical also diffuses into stagnant pores to be released slowly when the main body of chemical has passed. Because of this phenomenon, even soluble salts that do not adsorb on the soil tend to lag behind the movement of the solvent front.

The third factor of adsorption dynamics introduces a "frictional" effect, principally through slow release of chemical by the soil. Unless the water flow is extremely slow, there is a tendency for some of the chemical to trail behind the main body because the chemical cannot get off the soil fast enough. This will distort the basic chromatographic pattern and can, in fact, largely obscure it through retaining most of the material behind the chromatographic peak. A fourth factor is the pattern of rainfall and evapo-transpiration which determines the infiltration into and movement of water through the soil. Chemicals are leached downward following a rain, but later, as evaporation dries out the upper layers of soil, there is a compensating upward movement of water and chemical. Moreover, some of the rain runs off the soil surface and does not enter the soil. The result is that the net movement of a chemical downward is less than would be expected from the amount of precipitation, and the chemical is more diffused than would be expected from the net distance transversed.

ADSORPTION

Since adsorption is a principal factor in leaching, a number of laboratory tests have been developed which primarily measure the effect of adsorption on leaching. These include adsorption coefficient, soil thin layer chromatography, solvent-water partition coefficient and saturated soil columns. By largely ignoring factors other than adsorption, these tests for leaching are simplifications or idealizations from the "real world" situation. As such, they do not necessarily quantitatively measure the amount of leaching that would occur. They are, however, useful indices for comparing the leaching potential of one chemical with another.

The simplest index for leaching is the adsorption coefficient itself. Since chemicals with lower adsorption coefficients are leached to a greater degree, a ranking of a group of chemicals according to adsorption coefficients is therefore a ranking of their tendency to leach. Such a ranking made up from the compilation by Hamaker and Thompson (1972) is shown in Table 1 and is actually in good agreement with our general experience with leaching of pesticides. Chloramben and 2,4-D are, for example, at the correct end of the series for mobile chemicals, and the position of DDT and paraquat is in agreement with their almost complete immobility in soils. Other comparisons are likewise valid, as, for example, that simazine is more leachable than atrazine and monuron is leached more than diuron. The relative leachability of an unknown chemical can be estimated from this ranking. If a chemical has a K_{oc} value of 250, we can tentatively conclude that, other things being equal, it would leach about the same as chlorpropham and perhaps more than prometone.

The actual amount of leaching will vary from soil to soil, mainly in response to the organic carbon content. For this reason, the adsorption on the basis of organic carbon, K_{oc}, was used in Table 1.

$$K_{oc} = \frac{\mu g \text{ adsorbed/g soil organic carbon}}{\mu g \text{ dissolved/g solution}} \tag{1}$$

This adsorption coefficient has the virtue of being roughly independent of any particular soil. A more precise relationship between soil adsorption and other soil properties would be desirable but has not yet been found.

Another quantitative index for leaching is the R_f value from soil thin layer chromatography as developed by Helling and Turner (1968) and Helling (1971). The chemical is spotted onto soil-coated glass plates which are developed with water. Helling proposed that R_f values in a standard soil be used to define five

TABLE 1

Comparison of Adsorption Coefficients
for a Selected Group of Pesticides[1]

Chemical	K_{oc}[2]
Chloramben	12.8
2,4-D	32
Propham	51
Bromacil	71
Monuron	83
Simazine	135
Propazine	152
Dichlobenil	164
Atrazine	172
Chlorpropham	245
Prometone	300
Ametryne	380
Diuron	485
Prometryne	513
Chloroxuron	4,986
Paraquat	20,000
DDT	243,000

[1] Hamaker and Thompson (1972).

[2] $K_{oc} = \dfrac{\mu g \text{ adsorbed/g soil organic carbon}}{\mu g \text{ dissolved/gm solution}}$

mobility classes according to the following scheme:

Class:

1 (Immobile)	2	3	4	5 (Very Mobile)

R_f Range:

0 -0.09	0.1 -0.34	0.35-0.64	0.65-0.89	0.90-1.00

One disadvantage of this method is that these R_f ranges are for a
Hagerstown silty clay loam (2.42% organic matter) and will not
apply to R_f values for other soils with different organic matter
contents. Moreover, it is often difficult to detect the position

of a chemical on a soil plate and soil adsorption coefficients may be easier to obtain for some chemicals.

Simple chromatographic theory can be used to correlate adsorption coefficients with soil TLC R_f values and, thus, R_f values compared among different soils. If chromatographic movement through a soil column is treated according to the distillation theoretical plate theory (Block et al., 1958; Martin and Synge, 1941), a formula for R_f is obtained in terms of the relative cross-sectional areas of the liquid and solid phases and the partition of a chemical between solid and liquid phases.

$$R_f = \frac{A_L}{A_L + \alpha A_S} = \frac{1}{1 + \alpha(A_S/A_L)} \qquad (2)$$

Where A_S and A_L are cross-sectional areas of solid and liquid phases and α is the ratio of volume concentration in the solid phase to that in the liquid phase. For saturated conditions which will be assumed for a soil plate, $A_L + A_S = A$ (cross-sectional area), this can be written:

$$R_f = \frac{1}{1 + \alpha(A/A_L - 1)} \qquad (3)$$

When re-expressed in terms of the pore fraction of the soil (θ), density of soil solids (d_s), soil adsorption coefficient (K_{oc}), and percent organic carbon (% oc), this takes the following form:

$$R_f = \frac{1}{1 + (K_{oc})\ (\%\ oc/100)\ (d_s)\ (1/\theta^{2/3} - 1)} \qquad (4)$$

The ratio, A/A_L, is set equal to $1/\theta^{2/3}$ by analogy to the treatment of soil diffusion by Millington and Quirk (1961) where it serves to correct for the tortuousness of flow through the porous medium. In this case, it serves to relate the pore volume to the cross-sectional area of the liquid phase in a saturated soil.

The correlation of R_f and K_{oc} data by this formula is shown in Table 2. The measured R_f values are from Helling (1971) and K_{oc} from a compilation by Hamaker and Thompson (1972). Also shown are the mobility class ratings assigned by Helling and Turner (1968).

There is definite correlation between calculated and observed values for R_f, with the measured values being larger. This tendency would be expected from the fact that Helling made his measurements

TABLE 2

Prediction of Soil TLC R_f Values From Adsorption
Coefficients (K_{oc}) and Soil Properties in a
Hagerstown Silty Clay Loam
(θ = 0.5, d_s = 2.5 g/cc, % oc = % om/1.724 = 1.40)

Soil TLC R_f Value

Pesticide	K_{oc}[1]	Calc.[2]	Meas.[3]	R_f Meas. - R_f Calc.	Mobility Class[4]
Chloramben	12.8	.79	0.96	.12	5
2,4-D	32	.60	0.69	.09	4
Propham	51	.49	0.51	.02	3
Bromacil	71	.41	0.69	.28	4
Monuron	83	.37	0.48	.11	3
Simazine	135	.26	0.45	.19	3
Propazine	152	.24	0.41	.17	3
Dichlobenil	164	.23	0.22	-.01	2
Atrazine	172	.22	0.47	.25	3
Chlorpropham	245	.17	0.18	.01	2
Prometone	300	.14	0.60	.46	3
Ametryne	380	.11	0.44	.32	3
Diuron	485	.09	0.24	.15	2
Prometryne	513	.09	0.25	.16	2
Chloroxuron	4,986	.01	0.09	.08	1
Paraquat	20,000	.002	0.00	-.002	1
DDT	243,000	.0002	0.00	-.0002	1
			Ave. =	$\overline{.14 \pm .27}$	
				(95% Conf.)	

[1] Hamaker and Thompson (1972).

[2] $R_f = \dfrac{1}{1 + (K_{oc})\ (\% \text{ oc}/100)\ (d_s)\ (1/\theta^{2/3} - 1)}$.

[3] Helling (1971).

[4] Helling and Turner (1968).

θ = soil pore fraction
d_s = density of soil solid
om = organic matter
oc = organic carbon
K_{oc} = adsorption coefficient on the basis of organic carbon

from the front of the spot, while the calculations should give the position of the middle of the spot. Examination of Helling's paper (1971) suggests that this difference between measured and calculated R_f values is about 0.04 - 0.07, which is less than the average differences of 0.14 given in Table 2. However, there is such variability that this average is not different from zero at the 95% confidence level. This variability is probably due to uncertainty in K_{oc} drawn from investigations using many different soils, and a definitive test of the equation would require adsorption coefficients determined for the same soil--Hagerstown silty clay loam. This degree of correlation should, however, be sufficient for a rough classification of pesticides as to relative mobility.

Briggs (1973) has proposed relationships between the octanol/water distribution ratio of a chemical and both its adsorption coefficient and its R_f value in the same Hagerstown soil used by Helling. The following equations are linear regressions of data from four soils and 30 chemicals not actually specified:

$$\log K_{om} = 0.524 \ (\pm \ 0.048) \ \log P + 0.618 \ (\pm \ 0.113) \tag{5}$$

$$\log (1/R_f - 1) = 0.517 \ (\pm \ 0.022) \ \log P + 0.951 \ (\pm \ 0.075) \tag{6}$$

where,

K_{om} = Adsorption coefficient on soil organic matter basis

= K_{oc} x factor (factor is usually 1.724 but is not specified by Briggs)

R_f values in other soils for which soil adsorption coefficient, pore volume, solids density and organic carbon content are known can be estimated by means of equation 4.

Saturated soil columns represent another test for leaching potential that is predictable from adsorption, provided that the water is percolated slowly enough. The movement of water in soil columns is measured in terms of water entering the soil (R) rather than movement of the water front. The equation for this case is only a slight modification of the one proposed for estimation of R_f values for soil thin layer chromatography.

$$R = \frac{A}{A_L + \alpha A_S} = \frac{1}{A_L/A + \alpha(A_S/A)} \tag{7}$$

$$= \frac{1}{\theta^{2/3}(1 - K_{oc}) \ (\% \ oc/100) \ (d_s) + (K_{oc}) \ (\% \ oc/100) \ d_s} \tag{8}$$

Values of R calculated from the work of Davidson and Chang (1972),
Davidson et al. (1968) and Davidson and McDougal (1973) and
Huggenberger et al. (1972) are shown in Table 3 together with
theoretical calculations, both by the approximate equation 8 and
the more sophisticated theory explained and used by the authors in
each case. The degree of agreement between the observed values and
the approximate formula for slow rates of flow (<0.1 in/hr) is
quite remarkable (better, in some cases, than the more sophisticated
theory). However, it should be pointed out that the approximate
equation gives only the point of maximum concentration, while the
more complete equation attempts the more difficult job of computing
the concentration of chemical at every point along the column.

ADSORPTION DYNAMICS

A further demonstration of the effect of rapid flow on leaching
is work by Green et al. (1968) shown in Figure 1. The majority
of the material emerges ahead of the expected position, but a
considerable portion also trails behind. A likely explanation for
the early emergence is that some of the adsorbing surface, i.e ,
organic matter, is buried inside small aggregates usually found in
soil, so that with too rapid a flow for diffusion equilibrium,
the soil will act as if it has a lower adsorptive capacity than its
organic carbon content indicates. Another cause may also contribute,
which is that the establishment of adsorption equilibrium may be
slow compared to the rate of flow of the solution.

The tendency of the chemical to "streak" or "tail" is apparently
mainly because desorption is slower than adsorption and is partially
irreversible. The equilibrium point achieved by desorbing soil
containing chemical is not the same as that reached by adsorbing
chemical from solution: more remains on the soil. This has been
demonstrated for a number of chemicals, as, for example, disulfoton
by Graham-Bryce (1967). Hornsby and Davidson (1973) and Swanson
and Dutt (1973) have been able to successfully model the leaching
of fluometron and atrazine, respectively, by incorporating this
factor.

A practical significance of this is that many column-leaching
studies will tend to show leaching to greater depths than should
occur in the real world. A common form of column leaching experiment is to place the chemical on top of the soil column, pour in
the total amount of water, and allow it to infiltrate the soil as
fast as it will and drain as far as it will, i.e., to field
capacity. Under these circumstances, the flow rate is initially
quite rapid so that, on the average, it may be greater than the
7.7 cm/hr (3 in/hr) shown in Figure 1 for the work of Green

TABLE 3

Observed and Calculated Leaching Rates of Pesticides
in Saturated Soil Columns

Pesticide	R[1] Observed	R[1] Approximate Calculated	R[1] Exact[2] Calculated	Infiltration Rate in/hr	Comments
Picloram (Davidson & Chang, 1972)	2.41	1.39	1.44	0.83	< 2 mm
	1.47	1.42	1.72	0.079	
	1.93	1.37	1.48	0.83	< 0.42 mm
	1.37	1.38	1.48	0.082	
(Davidson & McDougal, 1973)	1.47	1.38	1.48	0.084	
Fluometron (Davidson et al., 1968)	1.64	1.39	1.31	0.73	
(Davidson & McDougal, 1973)	1.14	1.14	--	0.081	
Lindane (Huggenberger et al., 1972)	0.253	0.289	0.133	0.028	Silt Loam
	0.287	0.267	0.200	0.030	
	0.096	0.080	0.170	0.019	Sandy Loam
	0.082	0.080	0.065	0.020	
Prometryne (Davidson & McDougal, 1973)	5.2	6.10	--	0.082	

[1] $R = \dfrac{\text{cm moved by chemical}}{\text{cm of } H_2O \text{ entering the soil}}$

[2] Calculated by the authors using differential equations presented.

et al. (1968). The results of such a column leaching experiment
will not accurately represent leaching under field conditions where
the average rate of water infiltration is much slower.

A feature of the "one-shot", unregulated column is that it is
allowed to drain to field capacity, i.e., to an unsaturated condition. In contrast, the laboratory leaching columns with constant
flow rate are saturated with water. Saturated columns are used
partially because it is much easier to regulate their flow rate,
but also because the theory of saturated leaching is much better
understood than that for unsaturated leaching. Green et al. (1968)
have compared leaching under saturated and unsaturated conditions
with the results shown in Table 4. It appears that leaching under
unsaturated conditions is less effective than leaching under
saturated conditions, even though the same amount of water passes
through the column. In actuality, this might be expected from
the fact that a higher proportion of the chemical would be adsorbed
under unsaturated conditions, due to the higher soil-to-water ratio,
so that the slower desorption processes would have a bigger influence. The unregulated, single addition column would seem to have two
two phases as indicated by the following representation:

"One-Shot" Column Leaching

Phase 1	Effect
Approximate saturated flow with decreasing flow rate	More rapid movement of concentration peak with considerable chemical trailing

Phase 2	Effect
Slow drainage to field capacity	Additional movement, probably less than expected from adsorption

The net effect would be in doubt because of the opposing
tendencies of the two phases of the process. However, the result
may be that some of the material will be further advanced than
would be expected from adsorption but that much would be trailing
behind the expected spot.

An important technical consideration is drainage of water in
the column. To get normal drainage of the soil column, it should
be in direct contact with a bed of soil or equivalent porous
material. If the column of soil is simply suspended by a support
such as a screen, drainage is greatly impeded by capillary action

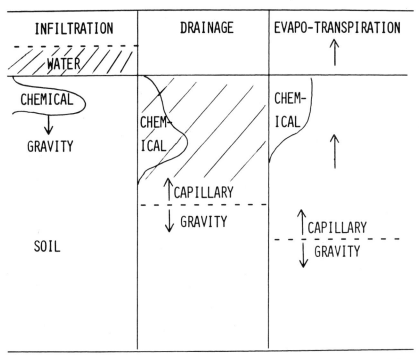

Figure 1. Comparison of measured atrazine elution with that predicted by a simple chromatographic model using the measured distribution coefficient of 3.0 (Green et al., 1968).

TABLE 4

Movement of Atrazine by Saturated
and Unsaturated Flow Through Soil[1]

Type of Flow	Flow Rate cm/hr	Water Content % Saturated	cm H_2O To Move Maximum Concentration To End of Column	cm Saturated/ cm Unsaturated
Saturated	0.6	100	12.7	1
Unsaturated	0.4	72	14.1	0.90
"	0.3	72	16.7	Average 0.83 / 0.76

[1] Green et al. (1968)

and the lower portions of the column may be in a saturated condition.
This is often overlooked, with the result that the leaching test
does not represent a normal field situation.

"REAL WORLD" LEACHING

Leaching under field conditions represents a further departure
from the ideal. Rain falls intermittently, followed by periods of
drying. Immediately after rainfall, water percolates downward,
leaching the chemical with it. However, after the rain, the
upper layers of soil dry out and capillary action brings water
from the deeper layers, carrying the chemicals upward   A representation of this process is shown in Figure 2.

The trailing of the band is shown as greater than that observed in the more ideal soil column leaching experiments. This
is partially because the chemical begins from a state of being
incorporated in the soil rather than from a state of being in
solution, and hysteresis, therefore, favors the chemical remaining
behind. An experimental example is shown by Table 5, calculated
from work by Burnside et al. (1963). Trichlorobenzoic acid was
applied to field plots of silty clay loam which were sampled to a
depth of 5 feet at 109 and 474 days after treatment. During the
474-day period, 38.68 inches of rain fell, which, if it all
entered the soil, would have saturated the soil to a depth of

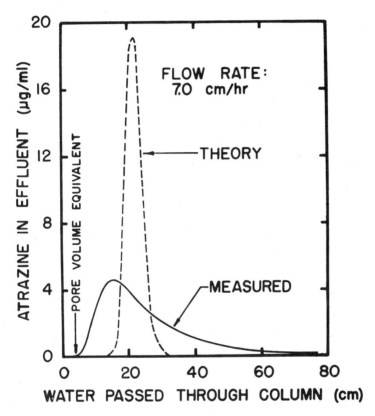

Figure 2.   Representation of leaching under field conditions.

about 9 feet and wet it to field capacity to about 23 feet.
The reported soil adsorption of trichlorobenzoic acid is very low,
i.e., nil (Hamaker and Thompson, 1972), so the chemical would have
moved with the water front in a soil column or other laboratory
leaching test. It is quite apparent that the movement of chemical
is far less than its negligible adsorption would suggest. Similar
results have been reported for other pesticides, such as for
picloram by Bauer et al. (1972) and dinoseb by Davis et al. (1954).
The marked retention in the top six inches of soil may be related
to the fact that this part of soil can dry out completely.

Leaching of fluometron under conditions of high rainfall and
a high water table was studied by La Fleur et al. (1973) with the
results summarized in Table 6. It will be noted that the movement
is not chromatographic, but that the same distribution pattern
remains even though 80% of the fluometron disappears from the
soil. On the average, over 70% remains in the top 23 cm (9 in)
with the rest distributed down to six feet. From an adsorption
coefficient (K_{oc}) of 31 and assuming 1% organic carbon, $\theta = 0.44$

TABLE 5

Movement of 2,3,5-TBA in a Field Plot of Wymore
Silty Clay Loam, Lincoln, Nebraska[1]

Depth Feet	109 Days		474 Days	
	Concentration Lbs/Acre-foot	Estimated % Remaining	Concentration Lbs/Acre-foot	Estimated % Remaining
0 -0.5	>>4	>>20	5	25
0.5-1	~4.68	23.4	0.5	2.5
1 -2	1.76	8.8	3.52	17.6
2 -3	1.60	8.0	0	0
3 -4	1.72	8.6	1.12	5.6
4 -5	0.92	4.6	1.12	5.6
		>73		56

[1] Burnside et al. (1963).

Total Rainfall: 38.68 inches
Maximum Water Front Movement: ~8.9 Feet for Saturation
(assume pore fraction = 0.4) ~22.5 Feet for Field Capacity
Application Rate: 20 lbs/acre foot

TABLE 6

Leaching of Fluometron Under Conditions of High Rainfall and High
Water Table in a South Carolina Coastal Plain[1]

Days Elapsed	cm Rain Since Last Date	Profile - Percent Recovered From Soil				Recovered From Soil (kg/ha)	Ground Water	
		0-23cm	23-45cm	45-81cm	81-182cm (~6 ft)		Conc (ppm)	kg/ha (Assume 1/2 Rainfall Percolates)
0		100				40[2]		
57	34.7	83	17	3	--	34.9	0.35	1.2
85	5.0	61	9	11	19%	38.0	0.15	.075
113	12.6	73	13	14	--	27.9	0.22	0.28
145	8.4	65	9	8	18	28.9	0.19	0.16
174	6.7	74	11	15	--	21.6	0.11	0.074
203	7.1	84	16	--	--	17.4	0.21	0.15
238	14.4	80	14	7	--	16.5	0.13	0.19
266	1.7	63	14	9	15	17.4	0.11	0.019
301	9.5	57	13	5	25	13.9	0.15	0.14
337	14.1	71	12	5	13	10.2	0.21	0.30
365	14.1	83	7	10	13	7.7	0.10	0.14
	128.3	(Ave)72.2	12.3	9.30	18.0		Total	2.73

1 La Fleur et al. (1973).

2 This level of application is 10 or more times commercial applications.

3 Position of water table approximately indicated by missing soil sample. Water table was from -41 to -192cm with an average of -91cm (3.0 ft).

and d_s = 2.5, a value of 1.1 for R is calculated by using equation
8. This means that the majority of the chemical should move 1.1
cm for each cm of water percolating through the soil; or, assuming
half the rainfall percolates to the water table, 70.9 cm for the
observed rainfall of 128.3 cm. The observed movement through the
soil is clearly less than this. The organic carbon content of the
soil was not provided, so a value of 1% was assumed. The uncertainty
of this assumption and the normal decrease in soil organic carbon
with depth may be an explanation for material found at greater
depths. Small concentrations of fluometron were found in the ground
water, and on the assumption that half the rainfall percolates to
the ground water, this represents 6.8% of the dose applied. Almost
half of this occurred in the first month, which may be due to the
fact that the well was not sealed from the surface layers of soil.
Presumably, the 74% not accounted for in the soil or water was
decomposed.

Further evidence that very little material actually leaches
through to the ground water is provided by the work of Edwards
and Glass (1971). Percolate was collected from lysimeters (8.07 m^2
x 2.44 m) to which 11.2 and 22.4 kg/ha of 2,4,5-T and methoxychlor,
respectively, had been applied. No methoxychlor was detected in
the percolate, and 0.5 ppb of 2,4,5-T appeared nine months after
application. At one year, 0.1 ppb was detected in two samples,
but in most samples, 2,4,5-T was not detectable.

In another study by the same authors, Glass and Edwards (1974)
the maximum concentration of picloram (1.2 ppb) appeared after
4700 liters of percolate. From the volume of the lysimeter,
assumptions of a pore volume of 0.4, and a field capacity of 36%
of saturation, the water content of the lysimeter is calculated
as 2797 liters, so the R_f value of the band appears to be 0.6,
which is much lower than the 0.96 value from soil thin layer
chromatography. Another important observation is that only 1.69 mg
of the original 1.8 g of picloram applied (0.093%) appeared in
2134 liters of percolate after a year. Dilution. decomposition,
and perhaps fixation are acting to severely limit the amount of
picloram reaching the bottom of the 2.44-meter lysimeter.

CONCLUSION

The field studies of leaching are difficult to analyze because
fundamental soil and physical properties are generally missing.
Further work is certainly needed, but available data indicate that
there is less movement of chemical than indicated by laboratory
leaching tests.

Although it is not yet possible to quantitatively estimate
leaching under field conditions for different soils and varying

climatic conditions, it is possible to use laboratory test results
to compare the degree of leaching of one compound with another
and, therefore, make use of available experience with leaching of
chemicals under field conditions. The amounts of chemicals
actually found to reach the ground water have been very small.
As pointed out initially, an understanding of the effect of
simultaneous leaching and degradation is needed to properly understand the movement of chemicals into the ground water.

SUMMARY

The leaching behavior of chemicals through soil depends upon
soil adsorption, hydrodynamic dispersion and diffusion, adsorption
dynamics, and evapo-transpiration. Most laboratory tests for
leaching depend upon adsorption modified by hydrodynamic dispersion
and diffusion, i.e., adsorption coefficient. octanol-water distribution ratio, soil thin layer chromatography and soil columns
run at low percolation rates. These do not quantitatively
predict leaching under field conditions, but can be used to compare
the leachability of one chemical with others whose leaching behavior
may be known. Leaching in soil columns, as usually conducted, tends
to overestimate the depth of penetration of chemical because of
rapid flow, but also overlooks the sizable fraction that trails
behind the peak. This trailing fraction which becomes the main
portion under field conditions, is apparently due to a slow rate
for desorption. Published studies show that little, if any, of
even quite mobile chemicals is leached to sufficient depth to enter
the ground water.

REFERENCES

Bauer, J.R., R.D. Baker, R.W. Bovey, and J.D. Smith. 1972.
    Concentration of picloram in the soil profile. Weed Sci.
    20:305.

Block, R.J., E.L. Durrum, and G. Zweig. 1958. A Manual of
    Paper Chromatography and Paper Electrophoresis. Second
    Edition. Academic Press, N.Y.

Briggs, G.G. 1973. A simple relationship between soil adsorption
    of organic chemicals and their octanol/water partition
    coefficients. Proceedings of 7th British Insecticide and
    Fungicide Conference.

Burnside, O.C., G.A. Wicks, and C.R. Fenster. 1963. The effect
    of rainfall and soil type on the disappearance of 2,3,6-TBA.
    Weeds 11:45.

Davidson, J.M., and R.K. Chang. 1972. Transport of picloram in
    relation to soil physical conditions and pore-water velocity.
    Soil Sci. Soc. Amer. Proc. 36:257.

Davidson, J.M., and J.R. McDougal. 1973. Experimental and
    predicted movement of three herbicides in a water-saturated
    soil. J. Environ. Quality 2:428.

Davidson, J.M., C.E. Rieck, and P.W. Santelmann. 1968. Influence
    of water flux and porous material on the movement of selected
    herbicides. Soil Sci. Soc. Amer. Proc. 32:629.

Davis, F.L., F.L. Selman, and D.E. Davis. 1954. Some factors
    affecting the behavior of dinitro herbicides in soil. Proc.
    Southwest Weed Conf. 7:205.

Edwards, W.M., and B.L. Glass. 1971. Methoxychlor and 2,4,5-T in
    lysimeter percolation and run-off water. Bull. Environ.
    Contam. and Toxicol. 6:81.

Glass, B.L., and W.M. Edwards. 1974. Picloram in lysimeter
    run-off and percolation water. Bull. Environ. Contam. and
    Toxicol. 11:109.

Graham-Bryce, I.J. 1967. Adsorption of disulfoton by soil. J.
    Sci. Fd. Agr. 18:73.

Green, R.E., V.K. Yamane, and S.R. Obien. 1968. Transport of
    atrazine in a latosolic soil in relation to adsorption,
    degradation and soil water variables. Trans. 9th International
    Soil Sci. Cong. 1:195.

Hamaker, J.W., and J.M. Thompson. 1972. Adsorption. In Organic Chemicals in the Soil Environment. C.A.I. Goring and J.W. Hamaker editors. Marcel Dekker, Inc., N.Y.

Helling, C.S. 1971. Pesticide mobility in soils, II: Applications of soil thin layer chromatography. Soil Sci. Soc. Amer. Proc. 35:737.

Helling, C.S., and B.C. Turner. 1968. Pesticide mobility: Determination by soil thin layer chromatography. Sci. 162:562.

Hornsby, A.G. and J.M. Davidson. 1973. Solution and adsorbed fluometron concentration distribution in a water-saturated soil: Experimental and predicted evaluation. Soil Sci. Soc. Amer. Proc. 37:823.

Huggenberger, F., J. Letey, and W.J. Farmer. 1972. Observed and calculated distribution of lindane in soil columns as influenced by water movement. Soil Sci. Soc. Amer. Proc. 36:544.

King, P.H., and P.L. McCarty. 1968. A chromatographic model for predicting pesticide migration in soil. Soil Sci. 106:248.

La Fleur, K.S., G.A. Wojeck, and W.R. McCaskill. 1973. Movement of toxaphene and fluometron through dunbar soil to underlying ground water. J. Environ. Qual. 2:515.

Lehav, N.J., and Y.Kahanovitch. Transport and persistence of pesticides in soil. Private communication.

Martin, A.J.P., and R.L.M. Synge. 1941. A new form of chromatogram employing two liquid phases. Biochem. J. 35:1358.

Millington, R.J., and J.P. Quirk. 1961. Permeability of porous solids. Trans. Faraday Soc. 57:1200.

Swanson, R.A., and G.R. Dutt. 1973. Chemical and physical processes of atrazine and distribution in soil systems. Soil Sci. Soc. Amer. Proc. 37:872.

PRINCIPLES OF PESTICIDE DEGRADATION IN SOIL

C.A.I. Goring and D.A. Laskowski
Ag-Organics Department
Dow Chemical Company
Midland, Michigan 48640

J.W. Hamaker and R.W. Meikle
Ag-Organics Department
Dow Chemical Company
Walnut Creek, California 94598

INTRODUCTION

History will record that eventually there was a relatively
simple accommodation to the extraordinarily complex problem of
defining and using principles of degradation to predict this aspect
of the behavior of pesticides in soil. So it is with many scientific endeavors.

What will be this accommodation? We have some ideas to present
but first we will review what is known about the subject. The
literature covering it is now so huge that it would be impossible
to discuss it in any kind of detail. We intend, instead, to
construct our subject matter from the views presented in a limited
number of papers and review articles that we believe represent
past and current thinking. We apologize for not being able to refer
to the individual contributions of the vast numbers of scientists
that have supplied the various pieces of this jigsaw puzzle.

BASIC CONSIDERATIONS

There are three important phases of pesticide degradation to
consider. These are: the causes of degradation; the pathways of
degradation; and the rates of degradation. Also it is important to
recognize that pesticides and their transformation products are a
result of both catabolic and anabolic processes in soil and that

135

we are really concerned with both kinds of processes and not just
those that are considered degradative.

Causes of Degradation

Three main types of transformation are reported to occur in
soil; photochemical, chemical and microbiological.

Although photodecomposition of pesticides in air and water is
a very common occurrence (Crosby, 1970; Crosby and Li, 1969; Helling
et al., 1971) it is doubtful that photodecomposition of pesticides
in soil has much practical significance. Few cases have been
reported and all are suspect because of the difficulty of excluding
the possibility of other pathways of decomposition or loss from soil
in the experiments conducted. In any case, little photodecomposition
in soil would be expected since radiant energy is strongly sorbed
by soil and therefore relatively unavailable for photochemical degradation even at the soil surface.

Chemical transformations in soil are widespread phenomena
(Crosby, 1970; Goring, 1967; Goring, 1972; Helling et al., 1971;
Kearney and Helling, 1969) that occur in several ways for many
types of pesticides as shown in Table 1. All of the reactions are
mediated through water functioning as a reaction medium, as reactant,
or both. Hydrolysis and oxidation are quite common phenomena.
Less is known about chemical reduction or isomerization. Nucleophilic substitution reactions other than hydrolysis may take place
with reactants dissolved in water, or with reacting groups of the
insoluble portions of the soil organic matter (Burchfield, 1960;
Crosby, 1970; Goring, 1967; Goring, 1972; Helling et al., 1971;
Kearney and Helling, 1969; Moje, 1960; Owens, 1963). Dismutations
of various types also occur and reactions with free radicals in
soil must be considered a distinct possibility (Crosby, 1970;
Goring, 1967; Helling et al., 1971; Steelink and Tollen, 1967).

The chemical reactions that occur in soil may be catalyzed
in several different ways (Crosby, 1970; Goring, 1967; Goring, 1972;
Helling et al., 1971; Kearney and Helling, 1969). Catalysis by
clay surfaces, metal oxides, metal ions, organic surfaces, and
organic materials that can be separated from the soil have all been
reported. Additionally, Skujins (1967) points out that extracellular enzymes are widely distributed and stabilized in soil.
Undoubtedly, they play a significant role in the degradation of
many pesticides and represent the transition point between
chemical and intracellular microbiological breakdown.

Microbiological breakdown of pesticides was viewed by Audus
(1960) as a process whereby microorganisms adapted to the pesticide
and produced enzymes suitable for degrading the pesticide during a

so-called "lag phase" followed by a period of "enrichment" as the adapted organisms multiplied while utilizing the substrate as a preferred energy source. He compiled a list of organisms shown to decompose certain herbicides including *Pseudomonas*, *Mycoplana*, *Agrobacterium*, *Corynebacterium*, *Arthrobacter*, *Flavobacterium*, *Achromabacter*, *Nocardia*, and *Trichoderma*. Audus did not consider these organisms catholic in their tastes since widely diverse structures could be utilized by the same genus of bacterium. As a matter of fact, he considered it most unlikely that any organic compound is completely immune to ultimate attack in soil by some organisms.

Unfortunately, his work led to a widespread belief that for microbiological breakdown of a pesticide to occur in soil there had to be specific organisms involved that could also be shown to accomplish this breakdown in pure culture using the chemical as a sole source of energy. The most extreme position was taken in a series of papers by Alexander (1965a, 1965b, 1966) in which he expounded the doctrines of "molecular recalcitrance" and "microbial fallibility". At that time he contended there were pesticides totally non-biodegradable in all biological systems (Alexander, 1966). Presumably this last conclusion was based on lack of demonstration that specific organisms could effect measurable changes in the molecules when used as a sole source of energy.

Many studies with soil have, however, indicated that there are many types of pesticides degraded in this medium even though chemical degradation was not known to occur and microorganisms that could utilize the pesticides as sole sources of energy had never been isolated. An outstanding example of this type of pesticide is picloram. Picloram is known to be degraded in soil by the microbial population and it is likely that populations of micro-organisms in soil vary in their capability to degrade the material (Youngson et al., 1967). Yet no microorganisms have as yet been identified that can use picloram as a sole source of energy. Never-theless, picloram is degraded by a wide variety of specific micro-organisms in pure culture providing they are supplied with a supplementary source of energy (Youngson et al., 1967). This type of degradation more recently emphasized by Alexander and Horvath as an important mechanism of pesticide breakdown in soil is now generally described as co-metabolism (Alexander, 1967; Horvath, 1971a; Horvath, 1971b; Horvath and Alexander, 1970). It probably occurs because the enzymes in the microorganisms are not highly specific for the substrates they ordinarily decompose and thus pesticides having some structural similarity to these substrates are decomposed without necessarily providing significant amounts of energy to the microorganisms. The diversity of microorganisms capable of degrading pesticides either by adaptation or co-metabolic processes is illustrated in Table 2. Certain types of organisms

TABLE 1

Chemical Reactions by Which Pesticides are Degraded in Soil[a]

Pesticide	Hydrolytic Nucleophilic Transformations	Chemical Reactions	
		Oxidation	Non-hydrolytic Nucleophilic Displacement Reactions
Abate		x	
Aldrin		x	
Amitrole		x	
Benomyl	x		
Binapicryl	x		
Captan	x		x
Carboxin		x	
Carboxylic Acid Esters	x		
Chloranil			x
2-Chloro-s-triazines	x		x
Chlorpyriphos	x		
Cycloate	x		
Dazomet	x		
Demeton	x	x	
Diazinon	x		
Dichlone	x		x
1,3-Dichloropropene	x		
Dichlorovos	x		

Difolatan	x		x
Dimethoate		x	
Dimethyldithiocarbamates		x	x
Dyrene	x		x
Endosulfan		x	x
EPTC			x
Ethylenebisdithiocarbamates		x	x
Fenflurazole			x
Formetanate			x
Imidan			x
Malathion			x
Metam-sodium		x	x
Methyl bromide	x		
Mevinphos			x
Nabam			x
Nitrohalobenzenes	x		
Phorate		x	
Phosphorothioic Acid Esters	x		x
Proxipham			x
Sesone			x
TEPP			x

[a] Burchfield (1960), Crosby (1970), Crosby (1973), Goring (1967), Helling et al. (1971), Kearney and Helling (1969), Moje (1960).

TABLE 2

Microorganisms Reported to Degrade Pesticides[a]

Microorganism	Pesticide
Achromobacter	Chlorpropham, 2,4-D, DDT, MCPA, 2,4,5-T
Aerobacter	DDT, endrin, methoxychlor
Agrobacterium	Chlorpropham, dalapon, DDT, picloram, TCA
Alcaligenes	Dalapon, maleic hydrazide, TCA
Alternaria	Dalapon
Arthrobacter	2,4-D, dalapon, diazinon, endothal, MCPA, picloram, simazine, TCA
Aspergillus	Atrazine, MMDD, 2,4-D, diphenamid, endrin, linuron, MCPA, monuron, PCNB, picloram, prometryne, simetryne, simazine, Telodrin*, trichlorfon
Bacillus	MMDD, dalapon, DDT, dieldrin, EPN, heptachlor, linuron, methyl parathion, monuron, parathion, picloram, Sumithion*, TCA
Bacteroides	Trifluralin
Botrytis	Picloram
Cephaloascus	PCP
Cephalosporium	Atrazine, prometryne, simetryne
Cladosporium	Atrazine, prometryne, simetryne
Clostridium	DDT, lindane, paraquat
Corynebacterium	2,4-D, dalapon, DDT, DNBP, DNOC, MCPA, paraquat
Erwinia	DDT,
Escherichia	Amitrole, DDT, lindane, prometryne
Flavobacterium	Chlorpropham, 2,4-D, dalapon, maleic hydrazine, MCPA, picloram, TCA
Fusarium	Aldrin, atrazine, DDT, heptachlor, PCNB, simazine, trichlorfon
Glomerella	PCNB, thiram
Helminthosporium	PCNB, picloram
Kurthia	DDT
Lachnospira	Trifluralin

Lipomyces	Paraquat
Micrococcus	Dalapon, TCA
Micromonospora	Heptachlor, TCA
Mucor	DDT, PCNB
Mycoplana	2,4-D, MCPA, 2,4,5-T
Myrothecium	PCNB
Neurospora	Chloroneb
Nocardia	Allyl alcohol, 2,4-D, 4-(2,4-DB), dalapon, DDT, heptachlor, PCNB, picloram, TCA
Penicillium	Aldrin, atrazine, MMDD, dalapon, heptachlor, monuron, PCNB, picloram, prometryne, propanil, simazine, Telodrin*, trichlorfon
Proteus	DDT
Pseudomonas	Allyl alcohol, chlorpropham, 2,4-D, dalapon, DDT, DDVP, diazinon, dieldrin, DNBP, DNOC, endrin, malathion, MCPA, monuron, PCP, phorate, simazine, TCA
Rhizoctonia	Chloroneb
Rhizopus	Atrazine, Dexon*, heptachlor
Saccharomyces	Captan, picloram
Sarcina	Monuron
Serratia	DDT
Sporocytophaga	2,4-D
Stachybotrys	Simazine
Streptococcus	DDT, heptachlor
Streptomyces	Dalapon, diazinon, PCNB, simazine
Thiobacillus	Phorate
Torulopsis	Phorate
Tramates	PCP
Trichoderma	Aldrin, allyl alcohol, atrazine, DDT, DDVP, diazinon, dieldrin, diphenamid, heptachlor, malathion, PCNB, PCP, picloram, simazine, TCA
Xanthomonas	Monuron

[a]Alexander (1969), Audus (1960), Helling et al. (1971), Menzie (1969), Woodcock (1967).
*Trademark

seem to predominate, specifically *Arthrobacter*, *Aspergillus*, *Bacillus*, *Corynebacterium*, *Flavobacterium*, *Fusarium*, *Nocardia*, *Penicillium*, *Pseudomonas*, and *Trichoderma*.

It is quite likely that the microbial population of soil is indeed "infallible' as far as degrading pesticides is concerned and that degradation in soil, however slowly, will always take place. Actually, the pesticides generally regarded as the most "recalcitrant" have half-disappearance times in soil of from one to ten years (Hamaker, 1972). The more stable fractions of the soil organic matter resist degradation for thousands of years (Alexander, 1965b). Thus in comparison with these fractions of the soil organic matter even the most persistent pesticides can scarcely be considered as unusually "recalcitrant".

Although degree of "recalcitrance" to decomposition is certainly associated with the inherent resistance of a structure to attack by microorganisms, a much more important factor may be the extent to which the material is available for degradation in the soil water. Thus materials that are highly insoluble in water or highly sorbed by the soil solids are apt to resist degradation (Furmidge and Osgerby, 1967) even though their inherent resistance to degradation may not be great. Presumably, extreme insolubility in water is a major reason for the great resistance of certain fractions of the soil organic matter to degradation.

Considerable effort is being spent trying to distinguish between chemical and microbiological degradation of pesticides in soil. The task is difficult because the transition from one to the other is neither abrupt nor well defined. Experimental attempts to separate chemical from microbiological degradation involving chemical, radiation, or heat methods of sterilization suffer from the handicap that the sterilization process may destroy not only the microorganisms but some of the other catalytic systems contributing to degradation. It has been suggested by Meikle et al. (1973) that widely different activation energies for microbiological (4-6 k cal/mole) versus chemical degradative processes (18-25 k cal/mole) could be the basis for determining which process is responsible for degradation. However, data on activation energies compiled by Hamaker (1972) do not suggest two distinct categories but values falling anywhere within a range of 1 to 32 k cal/mole. Presumably, this is because many of the compounds are simultaneously decomposed both chemically and microbiologically resulting in overall activation energies in soil somewhere between the expected extremes.

Pathways of Degradation

Studies directed to understanding the transformation pathways for pesticides in soil have mushroomed since radioactive tracers became available, and are extensively reviewed in a number of papers and books (Carter, 1969; Crosby, 1970; Crosby, 1973; Fang, 1969; Foy, 1969; Funderburk, 1969; Geissbuhler, 1969; Goring, 1967; Goring, 1972; Helling et al., 1971; Herrett, 1969; Jaworski, 1969; Kaufman, 1970; Kearney and Helling, 1969; Kearney and Plimmer, 1970; Knuesli et al., 1969; Loos, 1969; Meikle, 1972; Menzie, 1969; Probst and Tepe, 1969; Upchurch, 1972; Woodcock, 1967). Results for individual pesticides revealed various types of transformations achieved by both catabolic and anabolic reactions. As information accumulated attempts were made to classify these transformations. Kearney and Helling (1969), Kaufman (1970) and Crosby (1973) classified them according to general type of reaction. Meikle (1972) has emphasized the various kinds of transformation that could occur with specific functional groups choosing examples not only from the literature on pesticides in soil but also from the literature on biochemical transformation of all types of chemicals, especially by microorganisms. He wanted to focus on the concept that pesticides are not a special breed of chemicals and that they are degraded in the same ways as are naturally occurring chemicals.

The various types of chemical and microbiological pesticide transformations known to occur in soils and plants are shown in Table 3. Transformations in plants are included since soil includes the root systems of plants. No attempt has been made to show specific pesticide structures or metabolites since we wished to emphasize the types of groups that are being transformed rather than the specific metabolites created. However, examples are cited of pesticides (or their metabolites) that have been shown to be involved in the various types of transformation whether chemical or microbiological.

It is interesting but not surprising to note that almost all transformations are accounted for by oxidative, reductive, hydrolytic and conjugative reactions, which are the kinds of reactions fundamental to all biological systems. At least 26 different types of oxidative transformations have been observed. Only about seven types of reductive transformations have as yet been shown to occur but as studies on the anaerobic degradation of pesticides are expanded, more kinds of reductive transformations will almost certainly be revealed. At least 14 different types of hydrolytic transformations have been demonstrated. A pesticide transformation of great importance is dehydrohalogenation since such a large percentage of pesticides contain halogens.

TABLE 3

Types of Pesticide Transformations
in Soil and/or Plants[a]

TYPE OF REACTION	INITIAL GROUP	REACTION PRODUCT	PESTICIDE (OR THEIR METABOLITES) INVOLVED
Oxidative	$-\overset{\mid}{\underset{\mid}{C}}-H$	$-\overset{\mid}{\underset{\mid}{C}}-OH$	Carbaryl, Carbofuran, DDT, Dieldrin, Landrin*, Propoxur
	$H-\overset{\mid}{\underset{\mid}{C}}-OH$	$-C=O$	Carbofuran, Chlorfenvinphos
	$H-\overset{\mid}{\underset{\cdot H}{C}}-OH$	$\overset{O}{\underset{\parallel}{}}\;-C-OH$	Sesone
	aromatic ring $-O-\overset{\mid}{\underset{\mid}{C}}-$	aromatic ring $-OH$	Chloroneb, 2,4-D, Dicamba, MCPA, Methoxychlor, Propoxur, 2,4,5-T
	$-CH = CH-$	$-CH\overset{O}{\diagup\diagdown}CH-$	Aldrin, Carbaryl, 2,4-D, Heptachlor, Isodrin, Picloram

		Pesticides
—CH=CH— (aromatic)	OH —C=CH—	Benomyl, Carbaryl, Carbyne, 2,4-D, Dicamba, Dichlobenil, Diphenamid, Diuron, MCPA, Propoxur
—CH₂CH₂—C(=O)OH	—C(=O)OH	4-(2,4-DB), MCPB
(benzene ring with OH, OH)	HO—C(=O) ... O=C—OH (ring)	2,4-D, MCPA
Cl —C=CH— Aromatic or Aliphatic	OH —C=CH—	2,4-D, CDEC
Cl H (ring structure)	OH H (ring structure) Cl	2,4-D
—CCl₃	—C(=O)OH	DDT

Oxidative

Reactant	Product	Compounds
$-\overset{\mid}{\underset{\mid}{C}}-Cl$	$-\overset{\mid}{\underset{\mid}{C}}-OH$	CDAA, Dalapon, DBCP, Heptachlor, TCA
$-\overset{\mid}{\underset{\mid}{N}}-$	$-\overset{\mid}{\underset{\mid}{N}} \rightarrow O$	Schradan
$-\overset{\mid}{\underset{\mid}{N}}-$	$-N{\overset{O}{\underset{O}{\diagup\!\!\!\diagdown}}}$	Diuron
$-NH_2$	$-N=N-$	Dicryl* , Karsil* , Propanil
$-\overset{\mid}{N}-\overset{\mid}{\underset{\mid}{C}}-$	$-NH$	Aminocarb, Atrazine, Carbaryl, Chloroxuron, Diphenamid, Diuron, Mexacarbate, Monuron, Paraquat, Simazine
RNH_2	ROH	2-Chloro-s-triazines
$-N{\overset{OCH_3}{\underset{CH_3}{\diagup\!\!\!\diagdown}}}$	$-NHCH_3$	Metobromuron
$\overset{O}{\underset{}{\diagdown}}N-\overset{\parallel}{C}-N\diagup$	$\diagup\!\!\diagup NH\diagdown$	Benomyl, Chloroxuron, Diuron, Fluometuron, Monuron

Reactant group	Product group	Compounds
$-\overset{\mid}{\underset{\mid}{P}}=S$	$-\overset{\mid}{\underset{\mid}{P}}-O$	Abate*, Carbophenothion, Dimethoate, Disulfoton, Fenitrothion, Fensulfothion, Malathion, Parathion, Phorate
$-S-$	$-\overset{O}{\overset{\|}{S}}-$	Abate*, Aldicarb, Carbophenothion, Carboxin, Demeton, Disulfoton, Fenthion, Mesurol* Phorate, Prometryne
$-\overset{O}{\overset{\|}{S}}-$	$-\overset{O}{\overset{\|}{\underset{\|}{S}}}\underset{O}{}-$	Abate*, Aldicarb, Carbophenothion, Demeton, Disulfoton, Endosulfan, Fensulfothion, Fenthion, Mesurol*, Phorate, Prometryne
$-SCH_3$	$-SH$	Prometryne
$-SH$	$-S-S-$	Prometryne
$\overset{O}{\overset{\|}{C}}-S-CH_3$	$\overset{O}{\diagdown}COH$	Prometryne

Reactant	Product	Examples
$\backslash N - \overset{\overset{\text{S}}{\|\|}}{C} - S -$	$\backslash N - \overset{\overset{\text{S}}{\|\|}}{C} - S - S - \overset{\overset{\text{S}}{\|\|}}{C} - N \backslash$	Dialkyldithiocarbamates
$C = O$	$-\overset{\text{H}}{\underset{\|}{C}} - OH$	Chlorfenvinphos
$-C \overset{\overset{\text{O}}{\|\|}}{\underset{\backslash \text{OH}}{}}$	$- H$	2,4-D, DDT
$C = C$	$-\overset{\text{H}}{\underset{\|}{C}} - \overset{\text{H}}{\underset{\|}{C}} -$	DBCP, DDT
$- CCl$	$- CH$	DBCP, DDT, EDB, Heptachlor, Lindane, Methoxychlor
benzene ring with $-NO_2$	benzene ring with $-NH_2$	Benefin, Dinoben, DNOC, Fenitrothion, Parathion, PCNB, Trifluralin
$- N_3$	$- NH_2$	WL 9385
$\overset{\overset{\text{O}}{\|\|}}{S} -$	$- S -$	Phorate

Reductive

	Reactant group	Products	Examples
Hydrolytic (addition of water including oxidative hydrolysis)	$-C{\overset{O}{\underset{\\ }{\lVert}}}-O-C-$	$-C{\overset{O}{\lVert}}-OH + HO-C-$	Binapicryl, Butonate, Chlorthal-methyl, 2,4-D esters, Malathion
	$-C{\overset{O}{\lVert}}-N<$	$-C{\overset{O}{\lVert}}-OH + HN<$	CDAA, CIPC, Dichlobenil, Dicryl*, Dimethoate, Diphenamid, Imidan*, Karsil*, Propanil
	$-CCl_3$	$-C{\overset{O}{\lVert}}-OH + HCl$	Nitrapyrin
	$-C\equiv N$	$-C{\overset{O}{\lVert}}-NH_2$	Dichlobenil
	$-CH{\overset{O}{\underset{\\ }{\diagup\diagdown}}}CH-$	$-\overset{OH}{\underset{H}{C}}-\overset{\\ }{\underset{OH}{C}}-$	Chlorfenvinphos, Dieldrin
	$-CCl$	$-COH + HCl$	Chlorfenvinphos, 2-Chloro-s-triazines, 1,3-Dichloropropene

Hydrolytic (addition of water including oxidative hydrolysis)		
$\begin{array}{c} S/O \\ \parallel \\ R_1O\!-\!P\!-\!R_3 \\ \mid \\ OR_2 \end{array}$ $R_2 = \text{alkyl}$	$\begin{array}{c} S/O \\ \parallel \\ HO\!-\!P\!-\!R_3 \\ \mid \quad \backslash \\ OH \quad OH \end{array}$ $+ R_1\,OH \text{ and } R_2\,OH$	Phosphorothioic, Phosphoric, and Phosphonic Acid Esters
$\begin{array}{c} O \\ \parallel \quad \diagup NH\,R_2 \\ R_1O\!-\!P \\ \diagdown NH_{R_2} \end{array}$ $R_2 = \text{alkyl}$	$\begin{array}{c} O \\ \parallel \\ HO\ P\!-\!OH \ + \\ \mid \\ OH \end{array}$ $R_1\,OH + R_2NH_2$	Nellite*
$\begin{array}{c} O \\ \parallel \\ RO\!-\!C\!-\!N\diagdown \end{array}$	$ROH + CO_2$ $+ HN\diagdown$	Carbaryl, Carbofuran, Carbyne, CIPC, Fenflurazole, Formetanate, IPC, Mexacarbate, Proxipham
$\begin{array}{c} O \\ \parallel \\ =\!N\!-\!O\!-\!C\!-\!N\diagdown \end{array}$	$\diagup N\!OH \ + \ CO_2$ $+ HN\diagdown$	Aldicarb
$\begin{array}{c} S \\ \parallel \\ RS\!-\!C\!-\!N\diagdown \end{array}$	$ROH \ + \ CS_2 \ +$ $HN\diagdown$	Dimethyl Dithiocarbamates
$\begin{array}{c} O \\ \parallel \\ -S\!-\!C\!-\!N\diagdown \end{array}$	$-SH \ + \ CO_2$ $+ HN\diagdown$	EPTC

$RO-\overset{\overset{O}{\|}}{\underset{\underset{O}{\|}}{S}}-ONa$	$ROH + HO-\overset{\overset{O}{\|}}{\underset{\underset{O}{\|}}{S}}-ONa$	Sesone
$\overset{O=}{\underset{O=}{S}}-O-$	$\overset{O=}{\underset{O=}{S}}-OH + HO-$	Chlorfenson
$-\overset{\overset{O}{\|\|}}{C}-OH + HO-$	$-\overset{\overset{O}{\|\|}}{C}-O-$	2,4-D, Picloram, Dicamba
$-\overset{\overset{O}{\|\|}}{C}-OH + H_2N-$	$-\overset{\overset{O}{\|\|}}{C}-NH-$	2,4-D, Diuron, Nitrapyrin, Picloram
$-OH$	$-OCH_3$	2,4-D
$-\overset{\overset{O}{\diagdown}}{\underset{}{C}}-OH + HO-$	$-\overset{\overset{O}{\diagdown}}{\underset{}{C}}-O-$	Abate*, 2,4-D, Dicamba, Dichlobenil, Diphenamid, Landrin*
$-\overset{\overset{O}{\diagdown}}{\underset{}{C}}-OH + H_2N-$	$-\overset{\overset{O}{\diagdown}}{\underset{}{C}}-NH-$	Amiben, Amitrole, Pyrazon
$\overset{\diagup}{\underset{\diagdown}{N}}-\overset{\overset{S}{\|\|}}{C}-SH + HO-$	$\overset{\diagup}{\underset{\diagdown}{N}}-\overset{\overset{S}{\|\|}}{C}-S-$	Dimethyldithiocarbamates

Conjugative
(elimination of
water, HX
(x = halo) and
H

	$-CX + HS-$ → $-C-S-$	Anilazine, CDAA, 2-chloro-s-triazines, Chlorthalonil, Methyl Bromide
	$-CX + HN\diagdown$ → $-C-N\diagdown$	Anilazine, CDAA, 2-chloro-s-triazines, Methyl Bromide
	$-CX + HO-$ → $-C-O-$	CDAA, 2-chloro-s-triazines, Methyl Bromide
	$-NH_2 + H_3C-$ → $-NH-CH_2-$	Amitrole
Dehydrohalogenation	$CH-C-Cl$ → $-C=C- + HCl$	2-Chloroethylphosphonic acid, DBCP, DDT, EDB, Lindane, Methoxychlor
Isomerization	$R_1O-\overset{S}{\underset{}{P}}\diagup^{OR_2}_{OR_3}$ → $R_1S-\overset{O}{\underset{}{P}}\diagup^{OR_2}_{OR_3}$	Phosphorothioic acid esters

[a]Carter (1969), Crosby (1969, 1973), Fang (1969), Foy (1969), Funderburk (1969), Geissbuhler (1969), Helling et al. (1971), Herrett (1969), Jaworski (1969), Kaufman (1970), Kearney and Helling (1969), Kearney and Plimmer (1970), Knuesli et al. (1969), Loos (1969), Meikle (1972), Menzie (1969), Probst and Tepe (1969).

*Trademark.

Despite the many different specific transformations that could occur with each pesticide (or its metabolites) in soil, normal rates of application of most pesticides do not give rise to a large number of identifiable metabolites. In actuality, only the few metabolites formed in the largest quantities and which are most resistant to degradation, occur at concentrations sufficiently high to permit (and for that matter to warrant) identification for any reasons other than curiousity. The remainder are too low in concentration and too transitory in soil to be of significance.

The end products of pesticide transformations in soil are carbon dioxide, water, mineral salts, metabolites naturally occurring in soil, and humic substances. A diagrammatic representation of the processes involved is shown in Figure 1. Some of the processes are reversible while others are not. For example, non-polymeric metabolites of the pesticide arising from conjugative reactions between carboxyl, hydroxyl and amino groups may be reversibly degraded to the original reactants but all processes leading from unnaturally occurring substances to naturally occurring substances in soil are by definition irreversible.

For many aliphatic pesticides and some types of aromatic pesticides the preponderance of the carbon of the pesticide is quite rapidly converted to carbon dioxide and evolved from the soil. However, there is always conversion of some of the carbon to natural soil constituents including humic substances. Furthermore, some aromatic pesticides are metabolized to structures having ring systems heavily substituted with such functional groups as NH_2, OH and COOH. These structures may then be incorporated by polymerization, oxidation, and reduction reactions into humic substances.

A very large amount of research has been done on humic substances in soil (Bremner, 1965; Felbeck, 1971; Haworth, 1971; Jackson et al., 1972; Schnitzer and Kahn, 1972). Humic substances are usually divided into three main fractions: fulvic acid (FA) which is soluble in acid and base; humic acid (HA) which is soluble in base but not acid; and humin which is insoluble in both. The three fractions are structurally similar to each other but differ in molecular weights, ultimate analysis and functional group content. The FA fraction has a lower molecular weight but higher content of oxygen-containing functional groups per unit weight. It is probably made up largely of phenolic and benzene carboxylic acids joined by hydrogen bonds to form a polymeric structure of considerable stability (Schnitzer and Kahn, 1972). The HA and humin fractions have much higher molecular weights and probably consist of large aromatic polymers with characteristics determined principally by the physical properties of large

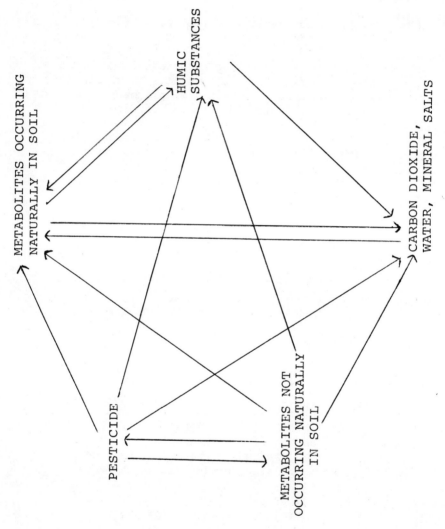

Figure 1. Pesticide transformation pathways in soil.

polymeric systems and the chemical properties of surface phenolic and carboxylic groups (Schnitzer and Kahn, 1972). Haworth (1971) believes that HA contains a complex aromatic core to which is attached, chemical or physically, polysaccharides, proteins, phenols and metals. The concept of an aromatic core is supported by the work of Jackson et al. (1972) which suggests a considerable amount of methylene bridging in a heavily branched three dimensional structure.

Less than half the nitrogen of humic substances is accounted for as amino acid nitrogen and hexoseamines (Bremner, 1965). Many other nitrogen containing compounds have been isolated and much of the nitrogen is thought to have been incorporated in the polymeric core rather than proteins (Bremner, 1965). Although some of the humic substance in soil is derived from altered lignin of plant material much of it is probably microbial in origin (Felbeck, 1971).

The complexity and diversity of the naturally occurring specific chemical entities that participate in the polymerization reactions responsible for the formation of humic substances is enormous (Bremner, 1965; Felbeck, 1971; Haworth, 1971; Jackson et al., 1972; Schnitzer and Kahn, 1972). On the other hand, the amounts of pesticide metabolites not naturally occurring in soil that are likely to be incorporated in humic substances will represent an extremely small portion of the total humic substances in soil. It is most unlikely that such metabolites can be released in their original forms from humic substance in which they have been incorporated (Schnitzer and Kahn, 1972). It is almost certain that together with naturally occurring humic substances they will be very slowly degraded in soil by microorganisms to CO_2 and water or very slowly transformed by ring cleavage into aliphatic compounds that readily serve as energy sources for the microorganisms (Schnitzer and Kahn, 1972).

Rates of Degradation

The quantitative aspects of pesticide degradation in soil have been discussed in considerable detail by Hamaker (1966, 1972) and Edwards (1972). Hamaker considered two basic types of rate laws. In the "power rate model" rate is proportional to some power of the concentration.

$$\text{Rate} = \frac{-dc}{dt} = kc^n \tag{1}$$

where c is the concentration, k is rate constant, and n is the order of the reaction.

In the "hyperbolic rate model" rate depends directly on the concentration and simultaneously on the sum of the concentration and other terms. In the simplest case these "other terms" are a single constant.

$$\text{Rate} = \frac{-dc}{dt} = \frac{k_1 c}{k_2 + c} \tag{2}$$

where k_1 is the maximum rate approached with increasing concentration and k_2 is a pseudoequilibrium constant (pseudo, because as the reaction occurs it is constantly unbalancing the equilibrium represented by the constant).

The "power rate model" is usually applied to chemical reactions in homogeneous solutions whereas the "hyperbolic rate model" is generally applicable to reactions catalyzed by adsorption to surfaces or complexing with catalyst molecules. Since for the most part degradation of pesticides in soil is catalyzed by intracellular enzymes, extracellular enzymes or other types of catalytic surfaces, the "hyperbolic rate model" is undoubtedly more applicable to pesticide degradation in soil than the "power rate model" on a theoretical basis. Indeed the "hyperbolic rate model" is a more general form of the Michaelis-Menten equation for enzyme kinetics shown below.

$$\frac{-dc}{dt} = \frac{V_m E c}{K_m + c} \tag{3}$$

where E is the concentration of enzyme, $V_m E$ corresponds to k_1 in equation (2) and K_m corresponds to k_2 in equation (2). However, the "power rate model" is a simpler more versatile tool for reaction rate modeling where the goal is to develop an empirical equation that best fits the degradation data obtained since it provides for reaction orders in excess of one.

It should be noted that for the "power rate model" when n is one, equation (1) reduces to equation (4).

$$\frac{-dc}{dt} = kc \tag{4}$$

which is the first order rate law. Likewise for the "hyperbolic rate model" when c is small compared to k_2, equation (2) also reduces to equation (4). For the "hyperbolic rate model" when c is large compared to k_2, equation (2) reduces to

$$\frac{-dc}{dt} = V_m E = k \text{ (for constant enzyme concentration)} \tag{5}$$

or zero order kinetics, i.e. the rate of degradation is independent
of concentration.

It is of interest to consider how the various types of degradative processes for pesticides in soil fit these rate laws.

The most complex process by which pesticides are degraded in
soils involves microbial utilization of the pesticide as an energy
source. Multiplication of the microorganisms involved occurs and
the rate of degradation increases both in proportion to their
numbers and any increase in efficiency due to adaptation by the
microorganisms. Eventually, multiplication slows down and finally
ceases due to the disappearance of the pesticide food supply. The
rate of degradation first ceases to increase and then decreases
approaching proportionality with the remaining pesticide concentration, i.e. first order. During the transition, the apparent
order will be somewhere between zero order and first order depending on the concentration. Because of the approximately linear
character of the early portion of any first order disappearance
curve a portion of the curve will appear approximately linear on
a c vs t graph. This type of decomposition typical of 2,4-D,
dalapon, pyrazon, is likely with chemicals easily degraded by
microorganisms.

The "co-metabolic" degradation curve in Figure 2 is typical
of the microbiological decomposition of pesticides which do not
serve as preferred energy sources for microorganisms and which are
moderately to weakly sorbed by soils. There is no lag period and
rate of degradation can shift from zero order at sufficiently
high concentrations to first order at low concentrations. Picloram
is an example of a pesticide degraded in this manner (Hamaker, 1972;
Hamaker et al., 1968; Meikle et al., 1973; Youngson et al., 1967).
Chemical degradation of poorly sorbed substances by catalytic
surfaces should follow a similar curve.

When pesticides are highly sorbed by the soil, the concentrations available for degradation are generally low relative to
the amounts of catalytic components in the soil whether decomposition is by microorganisms, extracellular enzymes, mineral
surfaces, or organic surfaces. The rate of degradation is initially
approximately first order as illustrated for the D_1 to E_1 portion
of the "co-metabolic" degradation curve.

However, the pesticide available for degradation is continually
being diminished by continued sorption and entrapment of the

pesticide in the soil organic matter matrix. Because of this the
rate of degradation decreases more rapidly than the total pesticide content would indicate which causes increased concavity of
the degradation curve as shown in Figure 2.

When the curves plotted on a linear scale as shown in Figure 2
are plotted on a logarithmic scale as shown in Figure 3, the types
of kinetics applicable to the various parts of the degradation
curves become more obvious. Both the curves for microbial adaptation and growth, and co-metabolism, are intially convex which is
typical of a rate of degradation less than first order. With
decreasing pesticide concentration they both reach straight line
status when the rate of degradation becomes first order. Entrapment of the pesticide in soil organic matter causes the degradation curve to become concave even when plotted on semi-log
paper suggesting a rate of degradation in excess of first order.
Actually it still represents first order degradation but the
rate constants continually decrease until the entrapment process
reaches a steady state. These curves are similar to logarithmic
curves previously described by Hamaker (1966) for various mathematical models of degradation.

Obviously, no one rate law or type of rate law is likely to
be completely adequate for any single pesticide over its entire
degradation curve, and even less so for a varied group of pesticides. Nevertheless, the use of first order constants to describe
pesticide degradation in soil has reasonable validity in a crude
way for many pesticides. For some pesticides, such as picloram,
the "hyperbolic rate model" shown in equation (2) is more applicable
than simple first order kinetics. This model is fairly adequately
represented by half-order kinetics as demonstrated for picloram
by Hamaker et al. (1968). A further refinement is the use of the
"power rate model" to establish the fractional order that best fits
the data obtained. Using this technique, Meikle et al. (1973)
found that a fractional order of 0.8 was most suitable for the
range of concentrations and the degradation data on picloram
obtained in their studies. Other types of empirical rate equations
have been proposed (Edwards, 1972; Hamaker, 1972) but it seems
doubtful that a single rate equation which is universally and
precisely applicable to all or even most pesticides will ever be
found for such a complex system as soil.

Nevertheless, rate equations are useful for describing approximately the rates of degradation of pesticides in soil. The first
order rate equation is most commonly employed largely because it
is the simplest equation to use. It has given rise to the "half-
life concept" which is the time required for half of the pesticide
to disappear from the soil, independent of the initial concentration
of the pesticide in the soil. However, degradation curves for

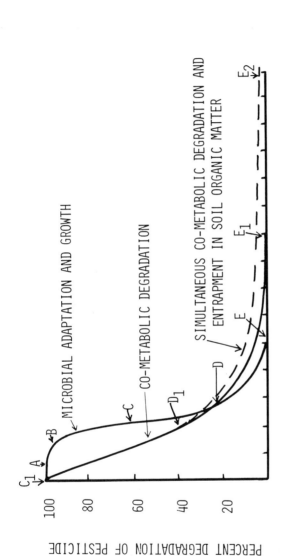

Figure 2. Degradation curves for pesticides in soil plotted on a linear scale. Microbial adaptation and growth: A → B lag period; B → C enrichment; C → D transition to first order rate of degradation; D → E first order rate of degradation to non-detectability. Co-metabolic degradation: C_1 → D_1 transition to first order rate of degradation; D_1 → E_1 first order rate of degradation to non-detectability. Simultaneous co-metabolic degradation and entrapment in soil organic matter: D_1 → E_2 entrapment in soil organic matter.

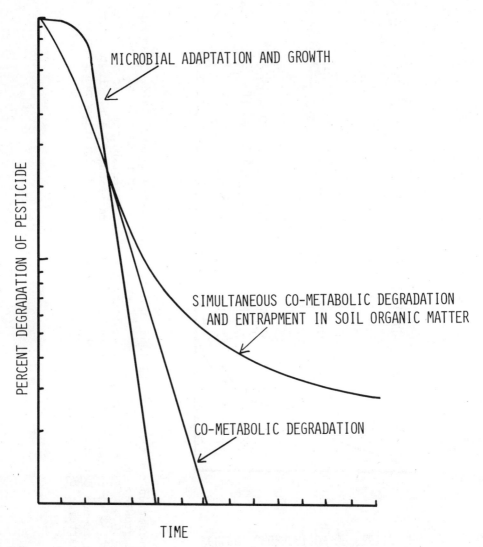

Figure 3.   Degradation curves for pesticides in soil plotted on a
logarithmic scale.

pesticides so frequently deviate substantially from first order
kinetics that Hamaker (1972) prefers the more laborious curve
fitting techniques and the use of the "power rate model". The
"power rate" equation derived can then be used to calculate an
instantaneous rate of degradation at any selected concentration or
time in the degradation process with greater precision than the
first order rate equation. Furthermore, it can be used to calculate
the time required for any percentage of degradation of the pesticide
at any specified concentration. Hamaker (1972) describes these
indices as DT_{50}, DT_{99}, etc. for 50% and 99% disappearance and
considers them useful for comparing rates of degradation of various
pesticides.

Factors Influencing Rates of Degradation

Attempts have been made to relate pesticide structure to susceptibility to decomposition but only in a limited way (Alexander,
1965b; Alexander, 1969; Kearney and Plimmer, 1970). Nevertheless,
a considerable number of useful generalizations can be made. Rates
of degradation of pesticides are reduced by sorption (Furmidge and
Osgerby, 1967) unless degradation is being catalyzed at the sorptive
soil surface. Thus highly polar non-ionic chemicals are likely to
be more easily degraded than non-polar non-ionic chemicals because
the latter are likely to be more water insoluble and highly sorbed
by the soil organic matter than the former. Similarly, anionic
chemicals are likely to be more easily degraded than cationic
chemicals because the latter are much more strongly sorbed by
soils. Aliphatic molecules or aliphatic portions of molecules are
generally more easily degraded than aromatic molecules or aromatic
portions of molecules.

Non-ionic molecules with active halogens are apt to be readily
degraded both chemically and microbiologically by nucleophilic
displacement reactions including hydrolysis (Burchfield, 1960;
Crosby, 1970; Corsby, 1973; Helling et al., 1971; Kearney and
Helling, 1969; Moje, 1960; Owens, 1963). Likewise ester linkages
separating highly polar groups are apt to be much more easily
hydrolyzed than ester linkages separating relatively non-polar
groups.

Compounds that are in a highly oxidized state may tend to
resist further oxidation and often represent poor sources of
energy for oxidative microbial attack under aerobic soil conditions.
The same compounds may be susceptible to reductive attack under
anaerobic conditions. For example, many of the chlorinated hydrocarbon insecticides are degraded very slowly in soil under aerobic
conditions but much more rapidly under anaerobic conditions
(Hamaker, 1972; Helling et al., 1971; Hill and McCarty, 1967).

Alternately compounds in a highly reduced state are more susceptible to decomposition under oxidative conditions than under
reducing conditions.

The positions of substituents on aromatic ring systems may
influence rates of degradation profoundly as pointed out by
Kearney and Plimmer (1970) and Alexander (1965b, 1969). In the
case of the chlorophenols substitution in the ring meta to the
hydroxyl confers much greater resistance to degradation than
substitution in the ortho or para position (Alexander, 1965b).

The principal climatic factors influencing rates of degradation of pesticides are temperature and moisture. Increasing
temperature will generally increase rates of degradation primarily
because of increased activity of the soil microorganisms. However,
increased temperature may also decrease sorption of the pesticide
making it more available for degradation in the soil solution.

Decreasing soil moisture generally decreases rates of degradation of pesticides under aerobic soil conditions. This is particularly true at lower soil moisture levels approaching air dry soil
conditions. The two principal reasons for the decreased rates of
degradation are reduced activity of the microorganisms and increased
sorption of the pesticide. However, in some instances where chemical decomposition is being catalyzed at the soil surface decreased
moisture may increase the rate of decomposition of the pesticide
(Goring, 1967).

At excessive moisture levels approaching saturation, the
soil may be switched from an aerobic to an anaerobic condition.
The rate of decomposition of the pesticide will then be determined
not by the moisture content of the soil but by the relative rates
of degradation of the pesticide under aerobic versus anaerobic
conditions.

The principal soil factors influencing rates of degradation of
pesticides are pH, organic matter, clay, iron and aluminum oxides,
and type of soil population.

Increasing pH will generally increase the rate of chemical
degradation of many pesticides (Goring, 1967; Helling et al., 1971)
especially the organophosphates. Increasing pH will also decrease
the sorption of anionic pesticides making them more available for
degradation.

Increasing soil organic matter may reduce the rates of degradation of pesticides in the soil solution because of increased
sorption. This is especially true of strongly sorbed pesticides.
However, for weakly sorbed pesticides the increased microbial

activity usually associated with increasing soil organic matter
could counterbalance the negative effect of sorption on decomposition of the pesticide. Furthermore, where chemical decomposition is being catalyzed at the organic matter surfaces, increased
organic matter is expected to increase the rate of decomposition
of the pesticide (Goring, 1967).

Clays, metallic cations such as iron and copper, and iron and
aluminum oxides notoriously catalyze the chemical degradation of
many pesticides in soil (Goring, 1967; Goring, 1972). However, it
is possible that microbiological degradation of anionic pesticides
such as picloram may be retarded by the presence of high amounts
of extractable aluminum (and possibly iron) in soils possibly
because of increased sorption of the pesticide (Meikle et al., 1973).

The soil factor having one of the most profound but unpredictable influences on decomposition of pesticides is the soil population. The nature and composition of the population is profoundly
influenced not only by climatic and soil factors such as temperature,
moisture, pH and organic matter but also by other factors such as
the mineral content, oxygen status, structure and porosity of the
soil. The constantly changing population will vary in its ability
to degrade the pesticide but in an unpredictable manner and will
always constitute a source of unpredictable variation in the rates
of degradation of pesticides in soil.

THE FIELD ENVIRONMENT

Reported information on the loss of pesticides from soils
under field conditions is relatively sparse and even fewer attempts
have been made to use this kind of information for the purpose of
predicting rates of loss under field conditions (Edwards, 1964;
Edwards, 1972; Hamaker, 1972). This is not too surprising considering the unmanaged complexity of the field environment in comparison with well controlled laboratory studies, and the limited
success achieved so far in being able to predict with great
accuracy rates of loss of pesticides from soil under controlled
laboratory conditions. Consequently, far too many investigators
and observers seem to have greater trust in their "intuitive
judgment" on the persistence of pesticides in soil than they do in
"quantitative prediction". Perhaps this is so because the extent
to which "quantitative prediction" misses the mark is easily
measured whereas there is no simple way of challenging the vagueness
of "intuitive judgment" and agreeing on the extent to which it has
failed.

The field situation is indeed far more complex than the controlled environments represented in many of the laboratory studies.

Edwards (1964, 1972) points out that in addition to loss as a result of degradation in soil, losses of pesticides under field conditions can take place during application, by leaching and run-off, by volatilization and by plant uptake. Furthermore, losses by leaching and volatilization are influenced by all of the soil and climatic factors that influence losses by degradation. Nor are conditions in the field either homogeneous or static. Temperatures, pH and moisture, organic matter and clay levels vary with soil depth. Cultivation and crop growth alters these relationships. Temperature and moisture fluctuations in the soil occur daily and rainfall patterns vary enormously.

In spite of the complexity of the field environment, the few attempts made at predicting losses of pesticides under field conditions have been surprisingly good. Dekker et al. (1965) used first order kinetics to predict the amounts of long-term residues that could be expected in 35 fields in Illinois that had been regularly treated with aldrin. The estimated amounts they expected in the fields, based on their premises were remarkably close to the amounts found by chemical analyses. Effects due to climatic variation were probably minimal in this study since all of the experimental plots were in Illinois.

Using a similar approach to data on the disappearance of aldrin reported by Lichtenstein and Schulz (1965), Hamaker (1966) was able to demonstrate quite close correspondence between residues of dieldrin in soil found and calculated. Climatological data was not used or needed in this study.

A study in which climatological data was essential has been reported by Hamaker et al. (1967). The disappearance of picloram from 200 different soil profiles in 18 different states and provinces of the U.S.A. and Canada was statistically correlated with temperature and rainfall. Soil factors, including pH, percentage organic matter, and clay, silt and sand distribution contributed little to the correlation. Temperature and rainfall, however, accounted for about half of the variability in loss of picloram. Four years later, Goring and Hamaker (1971) gleaned from the literature results from five additional studies on the loss of picloram under field conditions and calculated both observed and predicted half-order rate constants. The correlation coefficient between the observed and predicted constants was 0.87.

Only recently has a successful attempt been made to use laboratory data to predict rates of degradation under field conditions. Walker (1973) studied the effects of temperature and moisture on the rates of degradation of propyzamide in soil under laboratory conditions and developed a computer simulation model for predicting degradation. He then measured rates of degradation under field conditions. Using weather data as a measure of temperature and moisture conditions, rates of degradation predicted from his laboratory studies paralleled quite closely rates of degradation actually obtained under field conditions. This must be regarded as a landmark study since it has demonstrated the validity of predicting rates of degradation of pesticides under field conditions from measurements of rates of degradation under laboratory conditions.

THE BENCHMARK METHOD

We come now to what we believe is a relatively simple accommodation to the extraordinarily complex problem of predicting rates of degradation of pesticides under field conditions, namely the "benchmark" method. This approach has been briefly described by Hamaker (1972) for pesticide degradation in soil and by Goring (1972) for all aspects of pesticide behavior in the environment.

The sequence of events in the "benchmark" method is as follows:

1. Definition of the most important properties of pesticides that determine their behavior in the environment.

2. Development of laboratory methodology for measuring these properties.

3. Measurement of these properties for the major "benchmark" pesticides currently being used.

4. Establishment of the relationship between laboratory measurements and field behavior for "benchmark" pesticides.

The potential behavior of a new pesticide in the environment could then be predicted simply by measuring its properties and comparing the results obtained with those previously accumulated for "benchmark" chemicals.

The elegance of this approach is that it provides a rational matrix of information on the environmental behavior of pesticides that increases in scope, depth, and usefulness as each new chemical is incorporated into its proper place in the matrix.

Rate of degradation in soil is certainly one of the most important if not the most important characteristic of a pesticide determining its behavior in the environment. Laboratory methodology used to measure rates of degradation have varied widely making it extremely difficult to express the results obtained on a valid comparative basis. Most of the difficulty resides in the design

TABLE 4

Relative Persistence of Pesticides in Soils[a]

Pesticide Group	Pesticides	Estimated Time Required For 50% to Disappear From Soil (Months)
Non-persistent	Allyl alcohol, captan, chloropicrin, 2,4-D dazomet, DD, demeton, dichloran, dicrotophos, dimethoate, DNOC, formaldehyde, IPC, malathion, methyl bromide, Vapam*	<0.5
Slightly persistent	Aldicarb, amitrole, CDAA, CDEC, chloramben, chlorfenvinphos, chlorpropham, chlorpyrifos, CIPC, dalapon, diazinon, dicamba, disulfoton, DNBP, EPTC, fensulfothion, MCPA, methyl-parathion, nitrapyrin, parathion, phorate, propham, sesone, swep, 2,4,5-T, TCA, thionazin, vernolate	0.5 - 1.5
Moderately persistent	Atrazine, ametryne, bromacil, carbaryl, carbofuran chloroxuron, dichlobenil, diphenamid, diuron, Dyfonate*, fenthion, fenuron, ipazine, linuron, monuron, picloram, prometone, propazine, quintozene, simazine, TBA, terbacil, trifluralin	1.5 - 6
Persistent	Aldrin, BHC, chlordane, DDT, dieldrin, endrin, hexa-chlorobenzene, heptachlor, isodrin, lindane, methoxychlor, Telodrin*, toxaphene	>6

[a]Alexander (1965b), Alexander (1969), Crosby (1973), Edwards (1965), Edwards (1972), Graham-Bryce (1970), Hamaker (1972), Harris (1970), Helling et al. (1971).
*Trademark

of experiments. Investigators frequently select only one temperature, one moisture level, or one concentration of pesticide for
their experimentation. Their selection of soils may not span the
range of pH, organic matter and clay content likely to be encountered under field conditions. Furthermore, a sufficient number
of points to describe the degradation curves adequately may not be
obtained. In the absence of universal standardization of experimental parameters to be used, it is important that each investigator
include a range of values for his parameters such as moisture,
temperature, etc., that are sufficiently wide to encompass the
extremes to which the pesticide is likely to encounter in the
environment. In this way interpolation of the data between the
extremes will permit valid comparisons of results regardless of
the investigator and the pesticide involved.

Despite the paucity of adequate comparative data it is possible to crudely classify many pesticides with regard to their relative persistence in soil. Attempts at classification have
been made by Alexander (1965b, 1969), Graham-Bryce and Briggs (1970),
Edwards (1965, 1972), Crosby (1973), Helling et al. (1971), Harris
(1970), and Hamaker (1972). The latter is the most comprehensive
and quantitative study to date. Table 4 is a distillation of all
these studies.

It would be nice to achieve greater precision in the classification of the relative persistence of pesticides in soil than is
suggested by Table 4. Such a goal will not be accomplished with
the current haphazard approach to studying pesticide degradation
in soil. Some standardization of test methodology is needed before
substantial progress can be made.

SUMMARY

The causes of degradation of pesticides in soil including
chemical reaction, microbial enrichment, and co-metabolism are
discussed. The known pathways of degradation, including oxidative,
reductive, hydrolytic, conjugative, dehydrohalogenation, and
isomerization reactions, are summarized and their convergence
toward humification or complete degradation is described. The two
most important rate laws used to describe pesticide degradation in
soil, the "power rate model" and the "hyperbolic rate model", are
contrasted and the applicability of these rate laws to the various
types of degradative processes is considered with special emphasis
on the "half-life" concept. The effects of such factors as
chemical structure, temperature, water, oxygen, and various soil
properties such as organic matter, pH, and type of microbial population on rates of degradation are examined. The relationships
between laboratory and field degradation studies are considered,
and the "benchmark method" of predicting the persistence of
pesticides under field conditions is discussed.

REFERENCES

Alexander, M.   1965a.   Biodegradation; problems of molecular re-
    calcitrance and microbial fallibility.   Adv. Appl. Microbiol.
    7:35.

Alexander, M.   1965b.   Persistence and biological reactions of
    pesticides in soils.   Soil Sci. Soc. Amer. Proc. 29:1.

Alexander, M.   1966.   Biodegradation of Pesticides.   In Pesticides
    and Their Effects on Soils and Water.   ASA Special Publication
    Number 8.   p. 78.

Alexander, M.   1967.   The breakdown of pesticides in soils.   In
    Agriculture and the Quality of our Environment.   N.C. Brady,
    editor.   Am. Assoc. Adv. Sci. Publ. No. 85.   p. 331.

Alexander, M.   1969.   Microbial degradation and biological effects
    of pesticides in soil.   In Soil Biology Reviews of Research.
    UNESCO.   Vaillant - Cormanne, SA Liege, Belgium.   p. 209.

Audus, L.J.   1960.   Microbiological breakdown of herbicides in
    soils.   In Herbicides and the Soil.   E.K. Woodford and G.R.
    Sagar, editors.   Blackwell, Oxford.   p. 1.

Bremner, J.M.   1965.   Organic nitrogen in soils.   In Soil Nitrogen.
    W.V. Bartholomew and F.E. Clark, editors.   American Society of
    Agronomy, Wisconsin.   p. 93.

Burchfield, H.P.   1960.   Performance of fungicides on plants and
    in soil - physical, chemical, and biological considerations.
    In Plant Pathology, 3.   J.G. Horsfall and A.E. Dimond, editors.
    Academic Press, New York.   p. 477.

Carter, M. C.   1969.   Amitrole.   In Degradation of Herbicides.
    P.C. Kearney and D.D. Kaufman, editors.   Dekker, New York.
    p. 187.

Crosby, D.G.   1970.   The non-biological degradation of pesticides
    in soil.   In Pesticides in the Soil.   Intern. Symp.,Michigan
    State University.   p. 86.

Crosby, D.G.   1973.   The fate of pesticides in the environment.
    Ann. Rev. Plant Physiol. 24:467.

Crosby, D.G. and M. Li.   1969.   Herbicide photodecomposition.   In
    Degradation of Herbicides.   P.C. Kearney and D.D. Kaufman,
    editors.   Dekker, New York.   p. 321.

Dekker, G.C., W.N. Bruce, and J.H. Bigger. 1965. The accumulation
    and dissipation of residues resulting from the use of aldrin
    in soils. J. Econ. Entomol. 58:266.

Edwards, C.A. 1964. Factors affecting the persistence of insecti-
    cides in soils. Soils and Fertilizers 27:451.

Edwards, C.A. 1965. Effects of pesticide residues on soil
    invertebrates and plants. In Ecology and the Industrial
    Society. Fifth Symp. Brit. Ecological Soc. Blackwell,
    Oxford. p. 239.

Edwards, C.A. 1972. Insecticides. In Organic Chemicals in the
    Soil Environment. C.A.I. Goring and J.W. Hamaker, editors.
    Dekker, New York. p. 513.

Fang, S.C. 1969. Thiolcarbamates. In Degradation of Herbicides.
    P.C. Kearney and D.D. Kaufman, editors. Dekker, New York.
    p. 147.

Felbeck, G.T., Jr. 1971. Chemical and biological characterization
    of humic matter. In Soil Biochemistry, Vol. 2. A.D. McLaren
    and J. Skujins, editors. Dekker, New York. p. 36.

Foy, C.L. 1969. The chlorinated aliphatic acids. In Degradation
    of Herbicides. P.C. Kearney and D.D. Kaufman, editors.
    Dekker, New York. p. 207.

Funderburk, H.H. 1969. Diquat and paraquat. In Degradation of
    Herbicides. P.C. Kearney and D.D. Kaufman, editors. Dekker,
    New York. p. 283.

Furmidge, C.G.L. and J.M. Osgerby. 1967. Persistence of herbicides
    in soil. J. Sci. Fd. Agric. 18:269.

Geissbuhler, H. 1969. The substituted ureas. In Degradation of
    Herbicides. P.C. Kearney and D.D. Kaufman, editors. Dekker,
    New York. p. 79.

Goring, C.A.I. 1967. Physical aspects of soil in relation to the
    action of soil fungicides. Ann. Rev. Phytopathol. 5:285.

Goring, C.A.I. 1972a. Fumigants, fungicides and nematicides. In
    Organic Chemicals in the Soil Environment. C.A.I. Goring
    and J.W. Hamaker, editors. Dekker, New York. p. 569.

Goring, C.A.I. 1972b. Agricultural chemicals in the environment:
    A quantitative viewpoint. In Organic Chemicals in the Soil
    Environment. C.A.I. Goring and J.W. Hamaker, editors.
    Dekker, New York. p. 793.

Goring, C.A.I. and J.W. Hamaker. 1971. The degradation and movement of picloram in soil and water. Down to Earth 27(1):12.

Graham-Bryce, I.J. and G.G. Briggs. 1970. Pollution of soils.
R. I. C. Reviews 3:87.

Hamaker, J.W. 1966. Mathematical predictions of cumulative levels
of pesticides in soil. Adv. Chem. Series 60:122.

Hamaker, J.W. 1972. Decomposition: quantitative aspects. In
Organic Chemicals in the Soil Environment. C.A.I. Goring
and J.W. Hamaker, editors. Dekker, New York. p. 253.

Hamaker, J.W., C.R. Youngson, and C.A.I. Goring. 1967. Prediction
of the persistence and activity of TORDON herbicide in soils
under field conditions. Down to Earth 23(2):30.

Hamaker, J.W., C.R. Youngson, and C.A.I. Goring. 1968. Rate of
detoxification of 4-amino-3,5,6-trichloro-picolinic acid in
soil. Weed Res. 8:46.

Harris, C.R. 1970. Persistence and behavior of soil insecticides.
In Pesticides in the Soil. Intern. Symp., Michigan State
University. p. 58.

Haworth, R.D. 1971. The chemical nature of humic acid. Soil
Sci. 111:71.

Helling, C.S., P.C. Kearney, and M. Alexander. 1971. Behavior
of pesticides in soils. Adv. Agron. 23:147.

Herrett, R.A. 1969. Methyl- and phenylcarbamates. In Degradation
of Herbicides. P.C. Kearney and D.D. Kaufman, editors.
Dekker, New York. p. 113.

Hill, D.W. and P.L. McCarty. 1967. Anaerobic degradation of
selected chlorinated hydrocarbon pesticides. J. Water Poll.
Control. Fed. 39:1259.

Horvath, R.S. 1971a. Microbial cometabolism of 2,4,5-trichloro-
phenoxyacetic acid. Bull. Environ. Contam. Toxicol. 5:537.

Horvath, R.S. 1971b. Cometabolism of the herbicide 2,3,6-tri-
chlorobenzoate. J. Agr. Food Chem. 19:291.

Horvath, R.S. and M. Alexander. 1970. Cometabolism: A technique
for the accumulation of biochemical products. Can. J.
Microbiol. 15:1131.

Jackson, M.P., R.S. Swift, A.M. Rosner and J.R. Knox. 1972. Phenolic degradation of humic acid. Soil Sci. 114:75.

Jaworski, E.G. 1969. Chloroacetamides. In Degradation of Herbicides. P.C. Kearney and D.D. Kaufman, editors. Dekker, New York. p. 165.

Kaufman, D.D. 1970. Pesticide metabolism. In Pesticides in the Soil. Intern. Symp., Michigan State University. p. 73.

Kearney, P.C. and C.S. Helling. 1969. Reactions of pesticides in soils. Res. Rev. 25:25.

Kearney, P.C. and J.R. Plimmer. 1970. Relation of structure to pesticide decomposition. In Pesticides in the Soil. Intern. Symp., Michigan State University. p. 65.

Knuesli, E., D. Berrer, G. Dupuis and H. Esser. 1969. S-Triazines. In Degradation of Herbicides. P.C. Kearney and D.D. Kaufman, editors. Dekker, New York. p. 51.

Lichtenstein, E.P. and K.R. Schulz. 1965. Residues of aldrin and heptachlor in soils and their translocation into various crops. J. Agr. Food Chem. 13:57.

Loos, M.A. 1969. Phenoxyalkanoic acids. In Degradation of Herbicides. P.C. Kearney and D.D. Kaufman, editors. Dekker, New York. p. 1.

Meikle, R.W. 1972. Decomposition: qualitative relationships. In Organic Chemicals in the Soil Environment. C.A.I. Goring and J.W. Hamaker, editors. Dekker, New York. p. 145.

Meikle, R.W., C.R. Youngson, R.T. Hedlund, C.A.I. Goring, J.W. Hamaker and W.W. Addington. 1973. Measurement and prediction of picloram disappearance rates from soil. Weed Sci. 21:549.

Menzie, C.M. 1969. Metabolism of pesticides. Bur. Sport, Fisheries and Wild Life. Special Scientific Report - Wild Life. No. 127. pp. 1.

Moje, W. 1960. The chemistry and nematocidal activity of organic halides. Adv. Pest Control Res. 3:181.

Owens, R.G. 1963. Chemistry and physiolgoy of fungicidal action. Ann. Rev. Phytopathol. 1:77.

Probst, G.W. and J.B. Tepe. 1969. Trifluralin and related compounds. In Degradation of Herbicides. P.C. Kearney and D.D. Kaufman, editors. Dekker, New York. p. 255.

Schnitzer, M. and S.U. Kahn. 1972. Humic Substances in the
    Environment. Dekker, New York. pp. 137-201.

Skujins, J.J. 1967. Enzymes in soil. In Soil Biochemistry.
    A.D. McClaren and G.H. Peterson, editors. Dekker, New York.
    p. 371.

Steelink, C. and G. Tollen. 1967. Free radicals in soil. In
    Soil Biochemistry. A.D. McClaren and G.H. Peterson, editors.
    Dekker, New York. p. 443.

Upchurch, R.P. 1972. Herbicide and plant growth regulators. In
    Organic Chemicals in the Soil Environment. C.A.I. Goring and
    J.W. Hamaker, editors. Dekker, New York. p. 443.

Walker, A. 1973. Use of a simulation model to predict herbicide
    persistence in the field. Proc. Eur. Weed, Res. Coun. Symp.
    Herbicides-Soil. p. 240.

Woodcock, D. 1967. Microbiological detoxication and other trans-
    formations. In Fungicides, 1. D.C. Torgeson, editor.
    Academic Press, New York. p. 613.

Youngson, C.R., C.A.I. Goring, R.W. Meikle, H.H. Scott and J.D.
    Griffith. 1967. Factors influencing the decomposition of
    TORDON herbicide in soils. Down to Earth 23(2):3.

MODELING OF PESTICIDES IN THE AQUEOUS ENVIRONMENT

John P. Hassett and G. Fred Lee

Institute for Environmental Sciences

University of Texas-Dallas, Dallas, Texas 75080

INTRODUCTION

The United States Environmental Protection Agency has recently proposed water quality criteria for pesticides which are much stricter than any that have been applied before. For example, the U.S. EPA has proposed that the recommended maximum limit in whole (unfiltered) water for DDT be 0.002 ug/l. The recommendation for parathion and toxaphene are 0.004 ug/l and 0.01 ug/l, respectively. They have also recommended that general standard of 0.001 x 96 hr LC50, where 0.01 is the application factor which relates acute to chronic toxicity.

These criteria are based on acute and chronic toxicity of pesticides as well as accumulation of some of them in food webs, which can result in potentially significant concentrations in organisms of higher trophic levels. These new proposed criteria require a much better understanding of the environmental behavior of pesticides in order to predict their concentrations in aquatic systems and their impact on biological communities.

Pesticides manufacturers and environmental quality control regulatory agencies must develop a systematic approach for reviewing for each potential source the environmental behavior of all existing, and especially, new pesticides. This paper discusses the use of environmental chemistry models to predict the potential environmental behavior and impact of new pesticides.

PREVIOUS ATTEMPTS TO MODEL PESTICIDE BEHAVIOR

Two reports have been published recently of attempts to mathematically describe DDT behavior in the environment. These models focus on a pesticide already in the environment. However, a model that successfully predicts DDT behavior might be useful in predicting behavior of future pesticides.

The first model is one proposed by Harrison et al. (1970). This model is essentially an attempt to describe DDT behavior in a food web. Since it deals primarily with DDT in organisms, it is concerned with only a small fraction of the DDT present in the environment. Harrison et al. derived an expression for the flow of DDT through what they call the substrate, which includes soil, water, and sediment; however, it has no practical use since it does not include terms for volatilization, sorption-desorption, and microbial decomposition, the factors that are probably most important in controlling DDT movement and persistence.

The equations presented to describe DDT movement in a food web are based on the assumption that the sole source of DDT to a trophic level is that contained in material ingested by organisms of that trophic level, and that the means of elimination of DDT from a trophic level are death of an organism in that level, excretion, and metabolism. No consideration is given to the possibility that the rates of excretion and metabolism are dependent on the concentration of DDT in an organism. The basic limitation of this model is that it fails to take into account the possibility of partitioning of DDT between lipids in an organism and the surrounding water. Hamelink, Waybrandt, and Ball (1971) found that algae, crustaceans, and fish were able to accumulate DDT directly from water and, in the case of crustaceans, concentrations of DDT in the organisms declined when DDT concentrations in water were reduced. Johnson et al. (1971) have found that crustaceans concentrate DDT from water and, in the case of *Daphnia magna*, can accumulate levels of over 100,000 times the DDT concentration in water. Oysters are also quite effective at concentrating DDT and in at least one instance (Westlake and Gunther, 1966) removed DDT from water with a calculated efficiency of about 90 percent.

Although it appears that many invertebrates are very effective at concentrating DDT from water, this process may not be as important in fish. Macek and Korn (1970) compared rates of accumulation of DDT from water and from food by brook trout. They found that at concentrations to which the trout would be exposed in the environment, DDT was accumulated about ten times more rapidly from food than from the surrounding water. Thus, while direct exchange of DDT between organisms and water is probably very important for organisms in lower trophic levels, such as algae and crustaceans, it may be less important for higher trophic levels containing fish.

Harrison et al. (1970) make further assumptions to simplify
their equation and arrive at the conclusion that the time required
to reach an "equilibrium" concentration in a trophic structure
"lies between four times the average life span of the longest-lived
species and the sum of the life spans for all trophic levels," a
conclusion that has been widely quoted in the literature. A careful examination of some of the simplifying assumptions, that organisms retain all DDT ingested and that they neither metabolize
nor excrete DDT, indicates however, that the assumptions and thus
the conclusion are probably in error. Macek et al. (1970) found
that rainbow trout retained only 20 to 24 percent of ingested DDT,
and Macek and Korn (1970) found that brook trout retain about 35
percent after 120 days and that the amount retained decreased with
time. Thus, it appears unlikely that organisms retain all DDT that
they ingest. A number of aquatic organisms, including crustaceans
(Hamelink, Waybrandt, and Ball, 1971; Johnson et al., 1970) and
fish (Macek et al., 1970) are capable of metabolizing and/or excreting DDT. Even if the prediction of this model is correct, it
does not represent a significant advance since it predicts a time
limit for an "equilibrium" rather than what ultimate DDT levels
will occur in organisms.

The objective of Harrison et al., to reliably predict DDT
behavior in organisms, is a necessary goal for any attempt to model
the environmental impact of a pesticide. Unfortunately, they fall
far short of their objective and succeed instead in mathematically
obscuring the obvious, that some part of the DDT ingested by an
organism will be retained by it, while ignoring other vital factors
that control DDT distribution.

Another model of DDT movement has recently been proposed by
Woodwell, Craig, and Johnson (1971). This model is an attempt to
predict what will eventually happen to DDT in the environment on
a global scale. Their approach is to estimate residence times of
DDT in soil, in the atmosphere and in the upper, mixed layer of
the ocean, which they feel are the major DDT reservoirs. They then
use these to predict movement of DDT from soil to the atmosphere to
the upper layer of the ocean, and finally to the abyss.

After consideration of estimates of the amount of DDT that has
been applied to soils in the United States and the amount that is
remaining in these soils, Woodwell, Craig, and Johnson (1971) calculate a DDT soil residence time of about five years. They estimate
further that less than one percent of DDT is transported from the
point of application by harvest with the crop and about 0.1 percent
is lost in water runoff. They assume that biochemical degradation
is insignificant and that volatilization is the chief mode of removal from soil. Although it is possibly true that degradation
is insignificant compared to volatilization of DDT in soil, there
is no hard evidence to support this assumption. If degradation is

Harrison et al. (1970) make further assumptions to simplify their equation and arrive at the conclusion that the time required to reach an "equilibrium" concentration in a trophic structure "lies between four times the average life span of the longest-lived species and the sum of the life spans for all trophic levels," a conclusion that has been widely quoted in the literature. A careful examination of some of the simplifying assumptions, that organisms retain all DDT ingested and that they neither metabolize nor excrete DDT, indicates however, that the assumptions and thus the conclusion are probably in error. Macek et al. (1970) found that rainbow trout retained only 20 to 24 percent of ingested DDT, and Macek and Korn (1970) found that brook trout retain about 35 percent after 120 days and that the amount retained decreased with time. Thus, it appears unlikely that organisms retain all DDT that they ingest. A number of aquatic organisms, including crustaceans (Hamelink, Waybrandt, and Ball, 1971; Johnson et al., 1970) and fish (Macek et al., 1970) are capable of metabolizing and/or excreting DDT. Even if the prediction of this model is correct, it does not represent a significant advance since it predicts a time limit for an "equilibrium" rather than what ultimate DDT levels will occur in organisms.

The objective of Harrison et al., to reliably predict DDT behavior in organisms, is a necessary goal for any attempt to model the environmental impact of a pesticide. Unfortunately, they fall far short of their objective and succeed instead in mathematically obscuring the obvious, that some part of the DDT ingested by an organism will be retained by it, while ignoring other vital factors that control DDT distribution.

Another model of DDT movement has recently been proposed by Woodwell, Craig, and Johnson (1971). This model is an attempt to predict what will eventually happen to DDT in the environment on a global scale. Their approach is to estimate residence times of DDT in soil, in the atmosphere and in the upper, mixed layer of the ocean, which they feel are the major DDT reservoirs. They then use these to predict movement of DDT from soil to the atmosphere to the upper layer of the ocean, and finally to the abyss.

After consideration of estimates of the amount of DDT that has been applied to soils in the United States and the amount that is remaining in these soils, Woodwell, Craig, and Johnson (1971) calculate a DDT soil residence time of about five years. They estimate further that less than one percent of DDT is transported from the point of application by harvest with the crop and about 0.1 percent is lost in water runoff. They assume that biochemical degradation is insignificant and that volatilizatio is the chief mode of removal from soil. Although it is possibly true that degradation is insignificant compared to volatilization of DDT in soil, there is no hard evidence to support this assumption. If degradation is

significant, this would affect an estimate of the amount of DDT
entering the atmosphere but would have no effect on their estimate
of residence time in soil.

For atmospheric residence time, Woodwell, Craig, and Johnson
assume a time constant of about four years. They base this assumption on an estimated concentration in rain of 60 ng/l, which may
be too high, and an average annual rainfall of one meter. This
yields an atmospheric residence time of 3.3 years. They also base
it on the time constant for carbon dioxide transfer from the atmosphere to the ocean, which is about seven years. As they point
out in their paper, the estimated time constant of four years is
probably an upper limit. Photochemical degradation of DDT while
in the atmosphere, which is not considered in this estimate, may
be significant and could shorten the residence time considerably.

For the oceans the assumptions made are that the abyss is
infinite for the purposes of the model and that the rate of transfer of DDT from the upper mixed layer into the abyssal region ap-
proximates the rate of carbon dioxide transfer, which is about
four years. It is not known how closely DDT transfer follows
carbon dioxide transfer; however, there does not seem to be any
other data available.

Woodwell, Craig, and Johnson have used their model for two
sets of conditions. First, that DDT use will decline to zero by
1974, and second, that DDT use will increase. Assuming that use
declines to zero, the model predicts that DDT concentration reached
a peak in 1964, that DDT in the air reached a peak in 1966 and will
decline to ten percent of its peak concentration by 1984, and that
DDT in the mixed layer of the ocean reached a peak in 1971 and should
decline to ten percent of the peak value by 1993. Assuming that
DDT use increases rather than decreases, the concentrations in the
atmosphere and ocean should also increase.

Although this model is a step in the right direction in predicting environmental behavior of pesticides, there are a few problems with it. First, the approach taken was to base some rate
estimates on DDT concentrations measured in the environment. Obviously this approach cannot be used for a pesticide that has not
yet been introduced into the environment. Another problem is the
difficulty in making reliable estimates of DDT concentrations in
the atmosphere and ocean and in computing residence times. Thus,
although the mathematical equations may themselves be correct, the
numerical constants used to solve them may be seriously in error.
Perhaps the greatest problem with this model, however, is that it
predicts only average pesticide concentrations in the atmosphere
and oceans. As the designers of this model point out, local fluctuations in concentration may be expected to be large. In particular, areas such as lakes, rivers, and estuaries near points of

application would possibly be more affected than the ocean by surface runoff and might at times have undesirably high pesticide
concentrations not predicted by this model.

The United States Environmental Protection Agency is currently
developing a model for predicting pesticide movement from soil to
water (Nicholson, 1974). When completed, this model will be applied
on a river basin scale in order to establish safe limits for pesticide application. Such a model should be useful in predicting
localized impact of a pesticide on the environment.

FACTORS CONTROLLING ENVIRONMENTAL BEHAVIOR OF PESTICIDES

As indicated in the previous discussion, the currently available models of pesticide behavior have little or no predictive
value. An effective model must consider sources of a pesticide,
amount and rate of transfer through various parts of the environment, rates of physical, chemical, and biochemical transformations
of both the parent compound and any environmentally significant
degradation products, and the biological impact of the ultimate
environmental concentrations. Some of the factors that may be
significant in these processes are discussed here.

Pesticides may enter the environment through a number of different pathways (Westlake and Gunther, 1966), but the major sources
are probably agricultural use, industrial and municipal sewage, and
efforts to control aquatic weeds. Wastewater disposal and aquatic
weed control generally involve direct introduction of pesticides
into the aqueous environment, while pesticides applied for agricultural purposes must usually follow indirect routes to reach an
aquatic system.

The method of application partly determines the fate of an
agricultural pesticide. For example, Hindin, May, and Dunstan
(1966) found that less than 35 percent of DDT and other pesticides
applied by aircraft spraying reached the area being sprayed. Presumably, some of the remainder was carried into the atmosphere.
Judging from the concentration of organochlorine pesticides found
in rainwater in the United States (Cohen and Pinkerton, 1966) and
Britain (Tarrant and Tatton, 1968) atmospheric transport of pesticides is considerable and has even been proposed as the major
route by which DDT and related compounds reach the oceans (Risebrough, 1969; Woodwell, Craig, and Johnson, 1971). Other means
of applying pesticides, such as direct application of herbicides
to soil, probably result in less initial loss to the atmosphere.

Once a pesticide reaches the soil, a variety of fates may
befall it. If it is systemic some of it will be taken up by the

crop and either degraded, volatilized, or transported away with the harvest. Uptake of non-systemic pesticides such as DDT is small, and loss by this route is usually considered insignificant (Hindin, May, and Dunstan, 1966). Another possibility is that the pesticide may be carried away by seepage and enter the groundwater. This is especially possible for compounds that are fairly water soluble and are not strongly sorbed by soil particles. For example, 2,4-D in manufacturing wastes is able to enter groundwater and result in contamination of wells (Walker, 1961). Compounds that have a low water solubility and tend to be sorbed do not migrate significantly into the groundwater. For example, Terriere et al. (1966) estimated that less than 0.1 percent of DDT applied to orchards entered groundwater and found that most of the pesticide was still in the top ten inches of soil.

Another means of removal from the point of application is direct surface runoff. This can occur with the pesticide dissolved in the runoff water, associated with suspended material in the water or both. Although the percentage of the pesticide applied that is lost through this route may be small, Hindin et al. (1966) estimate that less than 0.01 percent of DDT applied to a cornfield was lost in the runoff, it probably represents the major pathway by which agricultural pesticides enter lakes, streams, and rivers. The amount entering the environment by this avenue would be dependent on the solubility of the pesticide, the suspended solids concentration in the water, the sorption characteristics of the soil, and the amount of precipitation and runoff. The factors responsible for loss of the bulk of a pesticide from soil are probably losses to the atmosphere and degradation by soil microorganisms. These routes may account for up to 50 percent of DDT lost from soil (Terriere et al., 1966).

Movement into the atmosphere may occur both by direct volatilization, which is probably enhanced by covaporization with water, and by association with particulate materials carried off by wind. Pesticides in the atmosphere may then reenter the aqueous environment in the form of dustfall and precipitation at another location. Biological degradation may be complete or it may simply result in new compounds, as is the case in the formation of DDE from DDT and dieldrin from aldrin. The environmental behavior and significance of possible metabolites must therefore also be considered. Pesticides exhibit a wide range of degradability, from a few weeks for organophosphorus pesticides such as parathion to several years for chlorinated hydrocarbons such as DDT and dieldrin (Kearney, Nash, and Isensee, 1969). The relative importance of volatility and degradability would depend on the characteristics of a pesticide. For example, for a slowly degraded, volatile compound such as DDT, volatilization may dominate (Woodwell, Craig, and Johnson, 1971) while for a rapidly degraded compound such as parathion, degradation most likely dominates.

Once a pesticide reaches the aquatic environment, either through runoff, direct application, or atmospheric fallout, it can undergo a number of interactions with other components of the system. Perhaps the most important factor is the interaction of the pesticide with suspended matter and sediment. The nature of this interaction will depend on solubility of the pesticide and characteristics of the sediment such as organic content, clay content, and pH. Natural organic matter may play a major role in sorption of pesticides onto sediments. Sediments with high organic contents show increased tendencies to sorb lindane (Lotse et al., 1968) and the herbicides 2,3-D, 2,4,5-T, and 4-amino-3,5,6,trichloropicolinic acid (Hamaker, Goring, Youngson, 1966). Humic material may be responsible for this phenomena since phthalate esters have been found associated with a soil humic fraction(Ogner and Schnitzer, 1970) and humic material has been found capable of strongly sorbing DDT from solution (Wershaw, Burcar, and Goldberg, 1969). Another factor that may influence sorption is pH. While pH seems to have little influence on sorption of chlorinated hydrocarbons (Huang, 1971), it does have a marked effect on pesticides containing acidic functional groups. Hamaker, Goring, and Youngson (1966) found that sorption of 4-amino-3,5,6-trichloropicolinic acid increased sharply below the pK of this compound and at any pH the amount sorbed was related to the amount of unionized acid present.

A pesticide that is associated with suspended material will eventually enter the sediment. Once in the sediment, it may be rereleased into the water, taken up by organisms, altered or degraded by microorganisms or simply buried. Many sediments are anaerobic, a condition under which many compounds are not readily degraded. Some compounds, however, such as DDT and lindane, are degraded more readily under anaerobic than aerobic conditions (Hill and McCarty, 1967). Thus, the migration of compounds such as DDT into sediments is desirable because it places them in an environment that promotes their degradation.

Some pesticides may remain dissolved in the water rather than enter the sediments, and pesticides in sediments will probably also be present in the overlying water, at least in trace quantities. Pesticides present in water may have a number of fates. They may be sorbed by the sediments, degraded by microorganisms, taken up by organisms or diluted in the oceans. Degradation by microorganisms is probably not as likely to occur in water as it is in soil or sediment because of the lower concentration of microorganisms. However, some degradation will probably occur. Uptake and concentration by organisms, while potentially important to the organisms themselves is very small compared to the total amount of pesticides in the environment, at least in the case of DDT (Woodwell, Craig, and Johnson, 1971).

EVALUATION OF BIOLOGICAL IMPACT

Once environmental pesticide levels have been predicted, it becomes necessary to determine their significance to aquatic organisms. The usual method for accomplishing this is to perform some sort of bioassay. This involves measuring either short-term acute effects or long-term chronic effects of the pesticide on a test organism. The organism chosen for the test should be the most sensitive one that is significant to the aquatic system in question, and the organism should be at the most susceptible stage of its life cycle.

A reliable bioassay must take into account chemical factors which may affect the availability of a pesticide to an organism. Because of the low water solubility of many pesticides, they are added to bioassay water by first dissolving them in acetone or some other organic solvent. This represents an unnatural situation which could invalidate bioassay results. Natural factors that may affect bioassay results include suspended solids and dissolved organic matter since a pesticide associated with either of these may be more or less toxic than the same pesticide in the free, dissolved state.

The problem of toxicity is not simply one of acute lethal effects; it also must include chronic, long-range effects such as reducing the ability of an organism to withstand stresses or interfering with its reproductive capacity. For pesticides that are biologically concentrated, the problem is further complicated because it is necessary to predict not only the effects of pesticide levels in water and sediment but also the effects of the ultimate concentrations in organisms.

SUMMARY

Proposed EPA water quality criteria will require a much better understanding of environmental behavior of pesticides. Transport-transformation models offer potentially useful tools for predicting environmental behavior; however, existing models for aquatic systems have little predictive value. Research is needed into specific factors that control environmental behavior of pesticides in order to develop more effective models. These models will become effective management tools that can be used to determine whether a new pesticide is likely to cause significant problems in the aquatic environment. It is proposed that a pesticide regulatory agency should require all manufacturers of pesticides to develop transport transformation models for all new pesticides and that these models must be developed by the agency prior to approval for large-scale manufacturing use of the pesticide. It is further recommended that the

pesticide regulatory agency require the pesticide manufacturer to
conduct the research necessary to determine how well the transport
transformation model actually predicted environmental behavior at
selected locations throughout the country. This requirement will
ultimately lead to a significant improvement in the ability to
model expected pesticide behavior, and thereby minimize the frequency with which national chemical crises such as caused by DDT
occur in the future. Also, the studies on the actual fate of the
pesticides used in the aquatic environment will provide the regulatory agency with the information needed to determine whether the
new pesticide could be having a significant deleterious effect on
aquatic ecosystems.

REFERENCES

Cohen, J.M., and C. Pinkerton. 1966. Widespread translocation of
      pesticides by air transport and rain-out. In: Organic Pesti-
      cides in the Environment. Amer. Chem. Soc. Adv. Chem. Ser.
      60:163-176.

Hamaker, J.W., C.A.I. Goring, and C.R. Youngson. 1966. Sorption
      and leaching of 4-amino-3,5,6-trichloropicolinic acid in soils.
      In: Organic Pesticides in the Environment. Amer. Chem. Soc.
      Adv. Chem. Ser. 60:23-37.

Hamelink, J.L., R.C. Waybrandt, and R.C. Ball. 1971. A proposal:
      exchange equilibria control the degree chlorinated hydrocarbons
      are biologically magnified in lentic environments. Trans. Amer.
      Fish. Soc. 100:207-214.

Harrison, H.L., O.L. Loucks, J.W. Mitchell, D.F. Parkhurst, C.R.
      Tracy, D.G. Watts, and V.J. Yannacone. 1970. Systems study
      of DDT transport. Science 170:503-508.

Hill, D.W., and P.C. McCarty. 1967. Anaerobic degradation of
      selected chlorinated hydrocarbon pesticides. J. Water. Poll.
      Cont. Fed. 39:1259-1277.

Hindin, E., D.S. May, and G.H. Dunstan. 1966. Distribution of
      insecticides sprayed by airplane on an irrigated corn plot.
      In: Organic Pesticides in the Environment. Amer. Chem. Soc.
      Adv. Chem. Ser. 60:132-145.

Huang, J. 1971. Effect of selected factors on pesticide sorption
      and desorption in the aquatic system. J. Water Poll. Cont.
      Fed. 43:1739-1748.

Johnson, B.T., C.R. Saunders, H.O. Saunders, and R.S. Campbell. 1971. Biological magnification and degradation of DDT and aldrin by freshwater invertebrates. J. Fish. Res. Bd. Canada 28:705-709.

Kearney, P.C., R.G. Nash, and A.R. Isensee. 1969. Persistance of pesticide residues in soils. In: Chemical Fallout. M.W. Miller and George G. Berg, editors. Charles C. Thomas, Springfield, Illinois, 54-67.

Lotse, E.G., D.A. Graetz, G. Chester, G.B. Lee, and L.E. Newland. 1968. Lindane adsorption by lake sediments. Env. Sci. Technol. 2:354-357.

Macek, K.J., and S. Korn. 1970. Significance of the food chain in DDT accumulation by fish. J. Fish. Res. Bd. Canada 27: 1496-1498.

Macek, K.J., C.R. Rodgers, D.L. Stalling, and S. Korn. 1970. The uptake, distribution, and elimination of dietary ^{14}C-DDT and ^{14}C-dieldrin in rainbow trout. Trans. Amer. Fish. Soc. 99: 689-695.

Nicholson, H.P. 1974. U.S. EPA Southeast Environmental Research Laboratory, Athens, Georgia. Personal communication.

Ogner, G., and M. Schnitzer. 1970. Humic substances: fulvic acid-dialkyl pthalate complexes and their role in pollution. Science 170:317-318.

Risebrough, R.W. 1969. Chlorinated hydrocarbons in marine ecosystems. In: Chemical Fallout. M.W. Miller and G.G. Berg, editors. Charles C. Thomas, Springfield, Illinois,5-23.

Tarrant, K.R., and J.O. Tatton. 1968. Organochlorinate pesticides in rainwater in the British Isles. Nature 219:725-727.

Terriere, L.C., U. Kiigemagi, R.W. Zwick, and P.H. Westigard. 1966. Persistence of pesticides in orchards and orchard soils. In: Organic Pesticides in the Environment. Amer. Chem. Soc. Adv. Chem. Ser. 60:263-270.

Walker, T.R. 1961. Groundwater contamination in the Rocky Mountain arsenal area, Denver, Colorado. Geol. Soc. Amer. Bull. 72: 489-494.

Wershaw, R.L., P.J. Burcar, and M.C. Goldberg. 1969. Interaction of pesticides with natural organic material. Env. Sci. Technol. 3:271-279.

Westlake, W.E., and F.A. Gunther. 1966. Occurrence and mode of
    introduction of pesticides in the environment. In: Organic
    Pesticides in the Environment. Amer. Chem. Soc. Adv. Chem.
    Ser. 60:110-121.

Woodwell, G.M., P.P. Craig, and H.A. Johnson. 1971. DDT in the
    biosphere: where does it go? Science 174:1101-1107.

ACCUMULATION OF CHEMICALS IN THE HYDROSPHERE

Rolf Hartung

Environmental and Industrial Health, School of Public

Health, University of Michigan, Ann Arbor, Michigan 48104

When considering accumulations of chemicals in the environment, among the first things that usually come into our minds are the concentrations and accumulations of various pesticides in aquatic food chains with subsequent detrimental effects on some species of birds and elevated levels of pesticides in fish which are to be used for food by humans. However, if one examines this story more closely, it becomes rapidly apparent that the behavior of persistent chemicals in the aquatic environment is considerably more complex than it appears at first sight. For the purposes of this symposium I shall limit my discussion to some members of the chlorinated hydrocarbon pesticides and PCBs, since these have been best studied with respect to their environmental behavior.

In considering the hydrosphere we must obviously consider far more than the water and those materials that are dissolved in it. The concept of the hydrosphere therefore includes important considerations relating to the lithosphere, the atmosphere, and the biosphere. It is obvious that these attempted partitions of the real world are highly artificial, and that considerable overlap exists among these spheres (Figure 1).

When water is analyzed for various chlorinated hydrocarbon pesticides it becomes immediately apparent that the levels to be found are exceedingly low, and that an unusual degree of variability exists (Table 1). Not all of this variability is due to the extraordinary experimental difficulties which are encountered. If one examines the concentrations of DDT in water in the Detroit River, one finds that these levels are relatively high and that they vary considerably with time (Figure 2). Some of this variability may have been due to the unrecognized presence of PCBs in the samples.

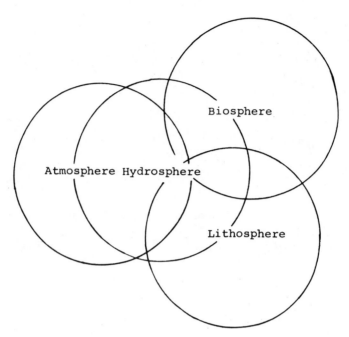

Figure 1.   Overlap among artificial subdivisions of the "real world."

TABLE 1

Concentrations of Chemicals in Water

Location	Chemical	Concentration	Reference
Lake Erie	p,p'DDT	4.8 ng/l	Hartung
Sargasso Sea	p,p'DDT	<0.15-0.5ng/l	Bidleman et al. 1974
Sargasso Sea	PCB	<0.9 -3.6ng/l	Bidleman et al. 1974
Lake Michigan	DDT	7 ng/l	Reinert 1970
Stream,			
New Brunswick	p,p'DDT	720 ng/l	Yule et al. 1971
Escambia Bay, Fla.	PCB	<100 ng/l	Duke et al. 1970
Escambia Bay	PCB	2.5 - 275µg/l	Duke et al. 1970
Irish Sea	PCB	<10 ng/l	Holdgate 1970
Irish Sea	DDT	<1-5 ng/l	Holdgate 1970
Hawaii (lakes)	p,p'DDT	5 ng/l	Bevenue et al. 1972
Hawaii (lakes)	dieldrin	1 ng/l	Bevenue et al. 1972
Hawaii (lakes)	lindane	5 ng/l	Bevenue et al. 1972

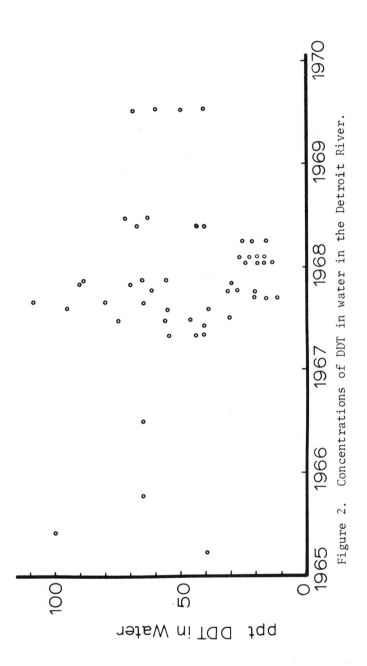

Figure 2.   Concentrations of DDT in water in the Detroit River.

However, if one scrutinizes the occurrence of high concentrations
of DDT in the water and relates these to climatic occurrences, one
finds that the highs appear to be related to rainfalls shortly
before the time the sample was taken. Therefore, some of these
elevations may have been due to runoff entering this water course
after rainfalls.

The DDT concentrations in the open lake environment of Lake
Erie are much less subject to variation. At a distance of approximately six miles offshore, they varied from 2.39 to 8.01 parts per
trillion over a sampling period of six months (Table 2). Reinert
(1970) also found concentrations of DDT of similar magnitude in
Lake Michigan. Fetterolf and Wilson (1969) showed that on the
average, over 50% of the dieldrin from unfiltered water samples
could be removed by filtration through a 0.6 micron membrane filter.

TABLE 2

p,p'DDT in Water and Plankton in Western Lake Erie

H_2O ppt	Plankton ppm (dry wt.)	Other ppm
3.21		
2.39		
3.99		Carp: 0.41 (wet wt.)
8.01	0.83	Mud: 0.73 (dry wt.)
5.19	0.45	0.92 (dry wt.)
6.75		
6.51		
5.30	0.79	Perch: 0.060 (wet wt.)
6.04	1.20	Common Shiner: 0.15 (wet wt.)
3.41		Mud: 0.10 (dry wt.)
4.47	0.45	
3.82		
3.62		
	0.80	
4.82 ± 0.46	0.75 ± 0.11	

At this point it must be remembered that nearly all of these analyses for DDT and water utilize whole water without filtration, so that no real differentiation is made between the amounts of DDT present on fine particles and the amount of DDT present in true solution. As a matter of fact it is probably almost impossible to define how much DDT is in true solution, because of adsorption onto container walls and contamination by plasticizers and other impurities during any cleanup process which will try to remove the particles from the water either by filtration or by high speed centrifugation. During a transect into Lake Erie it was found that during a moderately windy day, when wave action was stirring up bottom deposits near shore, the concentration of pesticide in water ranged from a high of approximately 50 ppt to a low of approximately 3 ppt about ten miles offshore (Table 3). At that point there was no further indication of turbidity due to wind action on the beaches. If indeed even a significant fraction of the remaining 3 ppt would be attached to particles, then the true concentration of DDT that is present in solution would be impossible to determine with present technology.

While it is difficult to determine the true environmental concentrations of DDT in the water column, it has also been difficult to determine the solubility of DDT in water. Babers (1955) found a solubility of 5.9 ppb DDT at 2°C, 37.4 ppb at 25°C, and 45 ppb at 37.5°. Bowman et al. found the solubility of C_{14} DDT to be 1.2 ppb or less at 25°C.

It is assumed that most of the present aquatic environment is not at saturation with respect to DDT, but it is clear that the quantities present in water at any one time are quite small, so that it is not very realistic to speak of the accumulation of pesticides in the water phase of the hydrosphere itself. Then where does any accumulation of chlorinated hydrocarbon pesticides occur? It appears that at any one time the largest quantity is present in the sediments. This does not imply that it reaches its highest concentrations within the sediments, those are reached within the fatty tissues of some aquatic organisms. However, the total mass of sediments which is able to concentrate chlorinated hydrocarbon pesticides far exceeds the biomass (Tables 1 and 4).

In addition there is evidence of some accumulation of pesticides at the air-water interface. I would like to try and explore the possible mechanisms involved in the concentration of pesticides at these various phases.

Pesticides occur in sediments in very much higher concentrations than they will occur in water itself. The concentration of pesticides in sediments is often considered to be due to adsorption. However, if one defines adsorption as the concentration of substances at interfaces, then it becomes apparent that the concentration of

TABLE 3

DDT Concentrations in Western Lake Erie Relative to
Distance From Shore (April, 1970)

Distance From Shore (km)	p,p'DDT in Water (ppt)
0.1	47
1.6	12
3.2	10
4.8	3.4
6.4	4.4
8.0	2.1
12	4.0
16	2.9

pesticides must also involve some additional processes. Most of
the chlorinated hydrocarbon pesticides are very non-polar substances. As such, one would not expect them to adsorb readily to
silicates or clays, because those would be surrounded with organized water molecules and an ionic double layer because of their
pronounced hydrophilic nature. These theoretical speculations
have been born out by experimental findings by Lotse et al. (1968).
He found that the adsorption of lindane into lake sediments was
greatly influenced by the occurrence of organic substances in the
sediments. Under natural conditions it is likely that many of the
hydrophilic surfaces would be covered by either ionic materials
or by surface-active agents. Other more hydrophobic surfaces, such
as carbon and sulfides, would be expected to attract organic materials which are present in greater abundance than the chlorinated
hydrocarbon pesticides. Such materials occur abundantly in the
environment. In addition to manmade oil pollution, there are also
many naturally occurring non-polar products, such as leaf epicuticular waxes, neutral hydrocarbons of recent origin, humates, and
fulvates. All of these have been identified in recent sediments.

Poirier and Thiel (1941) first proposed and demonstrated that
the sedimentation of oils in water could readily occur under laboratory conditions. Ludsack et al. (1957) found up to 17.8% sedimented oil downstream from a refinery in the Ottawa River in Ohio.
Hunt (1957) and Hartung and Klinger (1968) reported on the occurrence of sedimented oil in the Detroit River. North et al. (1964)
found sedimented oils after an oil pollution incident in a marine
cove in Baha California. Forbes and Richardson (1913) reported
2.6% oils in the bottom deposits of the Illinois River. McCauley
(1964) reported finding oily bottom deposits after oil pollution

Table 4

Concentration of Chemicals by Aquatic Organisms

Species	Chemical	Equilib. Conc. in Water (ppt)	Concentration Coefficient	Reference
Syracosphaera carterae	DDT	2.0	0.23×10^6	Cox 1970
Thalassiosira fluviatilis	DDT	2.6	0.29×10^6	Cox 1970
Amphidinium carteri	DDT	2.4	0.14×10^6	Cox 1970
carteri	DDT	3.3	0.16×10^6	Cox 1970
	DDT	1.0	0.22×10^6	Cox 1970
	DDT	0.5	0.14×10^6	Cox 1970
	DDT	0.8	0.12×10^6	Cox 1970
Chlorella sp.	DDT	100	0.34×10^6	Hartung
Chlorella reinhardi	DDT	220	0.03×10^6	Hartung
Selenastrum c.	DDT	220	0.34×10^6	Hartung
S. capricornutum	DDT	1000	0.10×10^6	Hartung
S. capricornutum	DDT	1500	0.09×10^3	Hartung
Ephemera danica	DDT	761	2.9×10^3	Södergren et al. 1973
Ephemera danica	PCB	526	2×10^3	Södergren et al. 1973
Daphnia magna	PCB	1100	47×10^3	Sanders et al. 1972
Gammarus pseudolim.	PCB	1600	27×10^3	Sanders et al. 1972
Orconectes nais	PCB	1200	5.1×10^3	Sanders et al. 1972

near Boston. Thus, while the reports may be scattered, the evidence
is clear that the existence of sedimented oils in association with
oil pollution is widespread.

In addition, there is an increasing body of evidence indicating
that aliphatic hydrocarbons are being synthesized by aquatic organisms and find their way into sediments in areas which have little or
no history of oil pollution (Han et al., 1968; Avigan and Blumer,
1968). Thus, hydrocarbons have been found in the recent sediments
of lakes in Minnesota (Swain, 1956) and the Gulf of Mexico (Stevens
et al., 1956).

Accumulations of waxy materials have been found in inlets in
British Columbia. The major constituents were found to be tetra-
decanoic acid and eicos-9-en-1-ol (Adams and Bonnett, 1969). It
is not clear whether the Bute Inlet waxes were of plant or animal
origin. But, among the constituents of leaf epicuticular waxes,
there occur high molecular weight organic chemicals which appear
to be resistant to degradation. These include alkanes such as n-
nonacosane, esters of long-chain alcohols with long-chain fatty
acids, such as n-hexacosanyl hexacosanoate, polycyclic hydrocarbons,
such as phyllocladene (Eglinton and Hamilton, 1967). Cooper and
Blumer (1968) identified fatty acids from C_8 to C_{21} in sediments,
many of which were of apparently recent origin.

We have particularly studied the relationship of petroleum
oils in sediments in the Detroit River. On the Detroit River it
has been shown that sedimented polluting oils can act as reservoirs
for chlorinated hydrocarbon pesticides. These sediments contained
0.1-2% of petroleum oils. The sedimentation of the oils is thought
to occur due to the coating of particles suspended in water and
subsequent sedimentation, as well as by sedimentation of oil droplets which have attracted airborne dust particles from water. There
is a significant correlation between the amounts of petroleum oil
in the sediments and the amount of DDT that is associated with it.
It is possible to determine the partition coefficient between oils
adsorbed onto sediments and then compare the partition coefficient
for oils and sediments with that between oils and water under laboratory conditions. It is found that the oil-water partition coefficient under both conditions is very similar (Hartung and
Klingler, 1970). In that study the partition coefficient for
p,p'DDT between water and free mineral oil was found to be 1.30 ±
0.26×10^6. For oily sediments from the lower Detroit River an
apparent partition coefficient of 1.45×10^6 was calculated from
monitoring data on DDT concentrations in water, oil concentrations
in sediments, and DDT concentrations in sediments.

In addition to the clearly organic chemicals in sediments
which have been cited above, one can also extract significant

quantities of sulfur from anaerobic sediments. It is not known to
what extent inorganic sulfur may bind or dissolve other organic
substances.

The classical methods of studying adsorptive phenomena may
not be able to clearly differentiate between adsorption and partitioning. Thus, partitioning of DDT between a single hydrophobic
solvent and water, could be readily described by a Lagmuir isotherm.
More complex systems involving adsorptive surfaces, several adsorbed
major solvents, and a solute such as DDT which may partition into
those solvents or adsorb directly, may produce other apparent adsorption isotherms.

In addition to partition phenomena in sediments, chlorinated
hydrocarbons also appear to accumulate in the surface areas of
water. According to Jarvis et al. (1967) most of the concentrated
surface microlayers, which are often noted as slicks, consist of
long-chain hydrocarbons which may contain waxes, fats, or petroleum products. Petroleum lumps have been found in neuston nets
with estimated concentrations of tars ranging from 1 to 20 milli-
grams per square meter (Morris, 1971). DDT and presumably other
chlorinated hydrocarbons have been shown to leave water by the
process of codistillation. The degree to which this is inhibited
by the presence of a micro-layer of non-polar substances is not
very well known. On the other hand the chlorinated hydrocarbon
pesticides have also been identified in atmospheric dust and in
rain water (Table 5). The concentrations found in rain water have
frequently been as high or higher than those found in fresh water.
It is likely that the elevated concentrations of pesticides found
in surface microlayers may be associated both with incoming rainfall as well as partitioning from water solution by the hydrophobic
substances in the surface microlayer. Data presented by Bidleman
and Olney (1974) show at least a two- to eight-fold enrichment in
the surface layers of Sargasso Sea water in comparison to subsur-
face water.

The uptake and storage of chlorinated hydrocarbon pesticides
by aquatic organisms is another important accumulative phase in
the hydrosphere (Table 4). Since this is also subject of another
paper in this symposium, I will report this phase only in general
outline. In the aquatic ecosystem one can also consider the uptake
of the chlorinated hydrocarbon pesticides in terms of partitioning
phenomena. The pesticides are not actively transported into the
organisms, but are incorporated through passive diffusion. The
primary driving force that is known to increase the concentration
of these pesticides from water into the living organisms is the
high partition coefficient between lipids and water. The cellular
membrane itself is composed of proteins and phospholipids and it
has been well established that materials which are lipid-soluble
find easier entrance into the cell. Within the organism the

TABLE 5

Chemicals in Rain and Snow

Location	Chemical	Concentration	Reference
Antarctic (snow)	DDT	40 ng/l	Peterle 1969
Antarctic (snow)	DDT	40 ng/l	Peterle 1969
Hawaii (rain)	p,p'DDT	1-14 ng/l	Bevenue et al. 1972
Hawaii (rain)	dieldrin	1-97 ng/l	Bevenue et al. 1972
Hawaii (rain)	lindane	1 -9 ng/l	Bevenue et al. 1972
G Britain (rain)	p,p'DDT	10-130 ng/l	Tarrant et al. 1968
G Britain (rain)	p,p'DDE	3-56 ng/l	Tarrant et al. 1968
G Britain (rain)	dieldrin	1-40 ng/l	Tarrant et al. 1968

chlorinated hydrocarbon pesticides are almost exclusively associated with lipids. Thus according to Reinert (1970) when one considers the pesticide burdens in fish in the Great Lakes on a fat basis alone, one finds that they tend toward constant concentrations, and the apparent differences between herbivores and primary and secondary predators, which may hint at the existence of a food chain, actually appear to be fortuitous and are related more to the concentrations of fats in the bodies of herbivores versus carnivores. Reinert had previously shown that the uptake of pesticides by fish through gills can possibly be at least as important as that taken in by the diet. Ferguson and Goodyear (1967) reported that black bullheads were able to absorb enough endrin through the gills or perhaps through the skin under experimental conditions to cause mortality. Thus it becomes apparent that multiple routes of entry are available and that the routes and dynamics of uptake into the organisms are exceedingly complex. It does appear that every aquatic organism acts as a miniature separatory funnel straining these chlorinated hydrocarbon pesticides out of the solution. The degree of accumulation is dependent on the amount of water that is scrubbed, the solubility dynamics of the pesticide, the permeability of the cell membranes, and the rate of breakdown within the organisms. However, in spite of all these impediments to uptake, the apparent equilibrium in small organisms is achieved within a few days, and concentration factors in the neighborhood of several thousand-fold are not uncommon at all (Table 4). The concentration factors have been determined so far mostly in the laboratory and only rarely in the environment. The actual magnitude of the concentration factor in the environment can only be speculated at, because of the proportion of dissolved chlorinated hydrocarbon pesticides with respect to the proportion that is adsorbed onto particles is usually not known.

In conclusion, the degree of accumulation of persistent pesticides in the hydrosphere depends on the segment of the hydrosphere that is being examined. The accumulation in the water phase itself is minimal. However, it does appear that the water phase acts as an important transport medium which facilitates the uptake of the persistent pesticides by aquatic organisms and assists in the establishment of equilibria. Most of the persistent pesticides found in the hydrosphere are associated with sedimentary levels. In this case, the primary influencing factors appear to be the concentrations of organic matter, possibly those consisting of petroleum products, waxes, and other natural products. The concentrations reached in sediments are usually never higher than those found in terrestrial environments which have received repeated applications of pesticides without subsequent disturbance.

Chlorinated hydrocarbon pesticides also accumulate in the microlayers at the air-water interface. This particular area has only been poorly evaluated and the biological significance of the pesticides that have partitioned into surface microlayers is poorly understood.

The uptake into biological organisms is without a doubt the most important. The very high partition coefficients between water and lipids favor extremely high uptakes by aquatic organisms. When calculated on a fat basis, it appears that concentration factors in the neighborhood of 1,000,000 are not unusual. Therefore, even with very low concentrations of pesticides in water, significant levels of pesticides in biological organisms can be reached. The food chain effects appear to be somewhat minimal and according to Hamelink et al. (1971) influence the ease with which solution equilibria can be reached. When aquatic organisms with high lipid content serve as food sources for terrestrial organisms, food chain effects can become predominant and allow further concentrations to occur.

Throughout the hydrosphere, partitioning phenomena appear to have a prominent role, and at times may entirely overshadow adsorption phenomena.

SUMMARY

The solubility of most of the environmentally important pesticides and other chemicals in pure water is remarkably low. In natural waters, only a minor portion of these pesticides is probably in true solution. The remainder is adsorbed onto particles or is partitioned into suspended organic solids or liquids. In spite of the small amount of pesticides found in true solution, the most pronounced examples of biomagnification occur in the hydrosphere.

Any logical consideration of the accumulation of chemicals in the hydrosphere must consider the total system, rather than only the phenomena that occur in water itself. And evaluation of the total concept of the hydrosphere indicates that the acutal accumulation of chemicals occurs in the sediments. The concentration of these chemicals in water is then determined by solubility, adsorption-desorption, partitioning, hydrodynamics, and other factors. Their concentration in the biota can be traced in part to the concentration found in true solution. It is not well understood to what extent pesticides attached to particles influence biomagnification. Future attempts at modeling and predicting the behavior of pesticides in the environment will need to take account of the complex basic phenomena which are active in the dispersal and accumulation of pesticides in the various segments of the hydrosphere.

REFERENCES

Adams, K.R., and R. Bonnett. 1969. Bute inlet wax. Nature 233: 943-944.

Avigan, J., and M. Blumer. 1968. On the origin of pristane in marine organisms. J. Lipid Res. 9:350-352.

Babers, F.H. 1955. The solubility of DDT in water determined radiometrically. J. Amer. Chem. Soc. 77:4666.

Bevenue, A., J.N. Ogata, and J.W. Hylin. 1972. Organochlorine pesticides in rainwater, Oahu, Hawaii, 1971-1972. Bull. Environ. Contam. Toxicol. 8(4):238-241.

Bidleman, T.F., and C.E. Olney. 1974. Chlorinated hydrocarbons in the Sargasso Sea atmosphere and surface water. Science 183:516-518.

Bowmann, M.C., F. Acree, Jr., and M.K. Corbett. 1960. Solubility of carbon-14 DDT in water. J. Agric. Food Chem. 8(5):406-408.

Cooper, W.J., and M. Blumer. 1968. Linear, iso and anteiso fatty acids in recent sediments of the North Atlantic. Deep-Sea Research 15:535-540.

Cox, J.L. 1970. Low ambient level uptake of ^{14}C-DDT by three species of marine phytoplankton. Bull Environ. Contam. Toxicol. 5(3):218-221.

Duke, T.W., T.I. Lowe, and A.J. Wilson, Jr. 1970. A polychlorinated biphenyl (Aroclor 1254) in the water, sediment, and biota of Escambia Bay, Florida. Bull. Environ. Contam. Toxicol. 5(2):171-180.

Eglinton, G., and R.J. Hamilton. 1967. Leaf epicuticular waxes. Science 156:1322-1335.

Ferguson, D.E., and P. Goodyear. 1967. The pathway of endrin entry in black bullheads. Copeia 2(5):467-468.

Fetterolf, C.M., Jr., and R.B. Willson. 1969. Pesticide runoff to Lake Michigan. Mich. Dept. Nat. Resources (mimeo) 11 pp.

Forbes, S.A., and R.E. Richardson. 1913. Studies on the biology of the upper Illinois River. Ill. State Lab. Nat. History Bull. 9:481-574.

Hamelink, J.L., R.C. Waybrant, and R.C. Ball. 1971. A proposal: exchange equilibria control the degree chlorinated hydrocarbons are biologically magnified in lentic environments. Trans. Amer. Fish. Soc. 100:207-214.

Han, J., E.D. McCarthy, W. VanHoeven, M. Calvin, and W.H. Bradley. 1968. Organic geochemical studies II. A preliminary report on the distribution of aliphatic hydrocarbons in algae, in bacteria, and in recent lake sediment. Proc. Natl. Acad. Sci. 59:29-33.

Hartung, R., and G.W. Klingler. 1968. Sedimentation of floating oils. Papers Mich. Acad. Sci., Arts and Letters 53:23-27.

Hartung, R., and G.W. Klingler. 1970. Concentration of DDT by sedimented polluting oils. Environ. Sci. Technol. 4:407-410.

Holdgate, M.W. 1970. Natural Environmental Research Council Report, London, quoted from N. Nelson et al. (1972). Polychlorinated biphenyls-environmental impact. Environmental Research 5:249-362.

Hunt, G.S. 1957. Causes of mortality among ducks wintering on the lower Detroit River. Ph.D. Thesis, The University of Michigan 296 pp.

Jarvis, N.L., W.D. Garrett, M.A. Scheiman, and C.O. Timmons. 1967. Surface chemical characterization of surface-active material in seawater. Limnol. Oceanogr. 12:88-96.

Lotse, E.G., D.A. Graetz, G. Chesters, G.B. Lee, and L.W. Newland. 1968. Lindane adsorption by lake sediments. Env. Sci. Technol. 2:353-357.

Ludsack, F.L., W.M. Ingram, and M.B. Ettinger. 1957. Characteristics of a stream composed of oil refinery and activated sludge effluents. Sewage Industr. Wastes 29:1177-1189.

McCauley, R.N.   1964.   The biological effects of oil pollution in
     a river.   Ph.D. Thesis, Cornell University, 173 pp.

Morris, B.F.   1971.   Petroleum: tar quantities floating in the
     northwestern Atlantic taken with a new quantitative neuston
     net.   Science 173:430-432.

North, W.J., M. Neushul, Jr., and K.A. Clendenning.   1964.   Succes-
     sive biological changes observed in a marine cove exposed to
     a large spillage of mineral oil.   Symp. on Pollution of the
     Sea by Microorganisms and Petroleum Products, Monaco, pp. 335-
     354.

Peterle, T.J.   1969.   DDT in Antarctic snow.   Nature 224:620.

Poirier, O.A., and G.A. Thiel.   1941.   Deposition of free oil by
     sediments settling in sea water.   Bull. Amer. Assoc. Petrol.
     Geol. 25:2170-2180.

Reinert, R.   1970.   Personal communication.

Sanders, H.O., and J.H. Chandler.   1972.   Biological magnification
     of a polychlorinated biphenyl (Aroclor 1254) from water by
     aquatic invertebrates.   Bull. Environ. Contam. Toxicol. 7(5):
     257-263.

Södergren, A., and Bj. Svenson.   1973.   Uptake and accumulation of
     DDT and PCB by Ephemera danica in continuous flow systems.
     Bull. Environ. Contam. Toxicol. 9(6):345-350.

Stevens, N.P., E.E. Bray, and E.D. Evans.   1956.   Hydrocarbons
     in sediments of the Gulf of Mexico.   Bull. Amer. Assoc.
     Petrol. Geol. 40:975-983.

Swain, F.M.   1956.   Stratigraphy of lake deposits in central and
     northern Minnesota.   Bull. Amer. Assoc. Petrol. Geol. 40:
     600-653.

Tarrant, K.R., and J.O'G. Tatton.   1968.   Organochlorine pesticides
     in rainwater in the British Isles.   Nature 219:725-727.

Yule, W.N. and A.D. Tomlin.   1971.   DDT in forest streams.   Bull.
     Environ. Contam. Toxicol. 5(6):479-488.

ESTIMATION OF SOIL PARATHION RESIDUES IN THE SAN JOAQUIN VALLEY,

CALIFORNIA - A SIMULATION STUDY

Dennis P.H. Hsieh, Haji M. Jameel, Raymond A. Fleck,
Wendell W. Kilgore, Ming Y. Li, and Ruth R. Painter
Food Protection and Toxicology Center
University of California
Davis, California 95616

INTRODUCTION

Despite the great concern over environmental contamination by
pesticides, man continues to depend on the use of pest-control
chemicals for agricultural productivity and disease control.
Phasing out of DDT has resulted in a considerable increase in
the use of organophosphorous insecticides, particularly parathion,
a chemical highly toxic for both invertebrates and vertebrates
including man. In 1970, 1.2 million lbs of parathion were used
in California, placing this compound among the top chemical pesticides in quantity used, and ahead of any other organophosphorous
insecticide.

Being a contact poison, parathion has an LD_{50} of ca 3 mg per
kg body weight for rats (Handbook of Toxicology, 1959), and is
generally considered to be a nonsystemic and nonpersistent insecticide. Its environmental impact and public health implications
therefore differ markedly from those of DDT, a much less toxic
yet far more persistent chemical, which has been used as a model
in most environmental impact studies. The need for further consideration of possible parathion residues in soils is emphasized
by the reports that parathion can penetrate and translocate
in plant tissues very rapidly (Kilgore et al., 1972), and that
when present in a relatively concentrated form, it undergoes slow
degradation in the environment (Hsieh et al., 1972; Wolfe et al.,
1973).

During the past three years, the University of California,
Davis, has been studying the fate of pesticides used in the eight

199

counties of the San Joaquin Valley. Many millions of lbs of
insecticides have been applied to agricultural areas in the valley
during the last two decades as it is one of the most intensively
cultivated regions in the world. This agricultural region has
been used as a model to study the fate of chemicals applied and
the ecological consequences of their use. As the first phase in
the systems approach, a simulation model was developed which
relates pesticide use patterns to residue levels in the soils of
the San Joaquin Valley. This report describes the simulation
model and the probable residue levels of parathion in the soils
of Fresno County, which is the most productive agricultural county
in the U.S., using the actual data of pesticide use for 1968 to
1972.

DATA BANK

The objective of the Data Bank is to logically and systema-
tically store the pesticide use data for simulation and other
applications. The original source of information is the Pesticide
Use Report, filed in the County Agricultural Commissioner's Office.
After preliminary checking, copies of the reports were forwarded
to the State Department of Agriculture, and then transferred to
magnetic computer tapes. Each record contains the day, month and
year of use; location (township, range, and section); acreage involved; commodity or crop treated; chemical and amount applied;
target pest; mode of application (aerial, ground, or other);
whether it is in combination with other pesticides; etc. Computer
programs were developed to print out not only the tabular read-outs,
but also the monthly-use charts (Figure 1) and the annual use-
density maps (Figure 2). A detailed description of the development
of the Data Bank will be reported elsewhere.

BACKGROUND OF SIMULATION

The terrestrial system of San Joaquin Valley is very varied,
and even the major agricultural areas comprise a large number of
parameters which interact to govern the fate of any chemical present
in them. It is not realistic, nor feasible, in a finite study
period to consider all possible interactions. Therefore, a simplified mathematical simulation was attempted.

Pesticide chemicals in the soil are primarily derived from
foliage sprays or dusts which miss their target and fall onto the
soil, either close to the target plants or at some distance after
drifting in air currents. Pesticide recoveries from aircraft
applications range from 15 to 99 percent of the amount applied,
primarily depending upon particle size of the spray and weather
conditions (Akesson et al., 1972). In addition, those chemicals

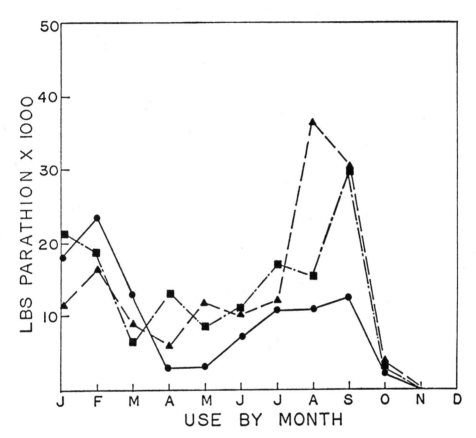

Figure 1.   Monthly use of parathion in Fresno County as reproduced from a computer output.   Symbols: ▲——▲ 1970;   ■——■ 1971;   ●——● 1972.

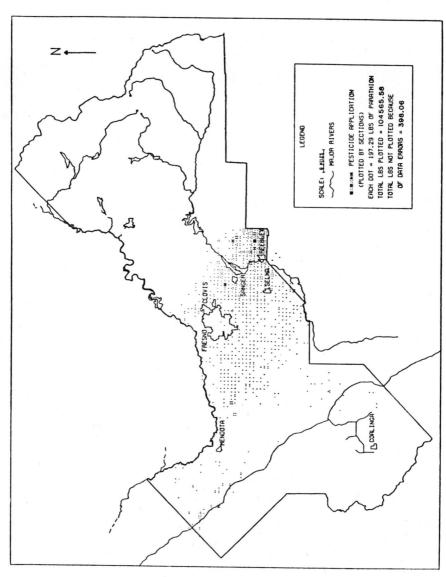

Figure 2. Density map of annual use of parathion in Fresno County in 1971.

originally retained on the foliage may reach the soil when they are washed or blown off plants, or when the plant residue is ploughed into the soil. Some chemicals are also applied by ground application to control soil-inhabiting pests.

The fate of a pesticide chemical in soil is governed by: 1) structure of the chemical as reflected in its volatility, solubility and reactivity; 2) concentration and formulation of the chemical as used; 3) properties of the surrounding soil including organic matter, moisture, clay, mineral ion contents, acidity, and temperatures; 4) cultivation patterns; and 5) rainfall, sunlight, and other weather conditions. The factors involved in the transformation and removal of pesticide chemicals in soils are so complex that a detailed process analysis is extremely difficult.

Edwards (1966) and Hamaker (1972), suggested a first order decay model for the fate of many pesticides in soil. The logarithmic disappearance of a chemical was proposed to involve four phases: 1) application losses due to sprays missing their target, drifting, and evaporation; 2) losses due to volatization from the soil surface; 3) losses resulting from leaching, adsorption and volatization after cultivation; and 4) losses resulting from chemical and enzymatic degradation, while leaching and volatization continue. The experimental evidence for the first order degradation assumption is available in the literature (Edwards, 1966; Hamaker, 1972; Iwata et al., 1973; Walker, 1973), which indicates that the rate of pesticide disappearance in soil is proportional to the residue level and a first order degradation rate constant.

SIMULATION MODEL

The model presented herein is a simplified case of the above assumptions. Losses due to the first three phases were combined into a single deterministic term defined as application loss (APLOSS). The amount of parathion used in each month, after correction for APLOSS (equation 1), was assumed to undergo first order degradation (equation 2). Thus, for any month, m:

$$A_m = B_m \times (1 - APLOSS) \tag{1}$$

$$R_m = A_m \times \exp(-0.5k) + R_{m-1} \times \exp(-k) \tag{2}$$

Where

A_m = net amount reaching soil in month m, lb.

B_m = total chemical used in month m, lb.

APLOSS = application loss according to mode of application, a fraction.

R_m = cumulative residue at the end of month m, lb.

k = overall degradation rate constant, $month^{-1}$.

R_{m-1} = cumulative residue before month m, lb.

In equation 2, the residue level prior to m, R_{m-1}, was subject to degradation for a full month, but the amount applied during the month, A_m, undergoes degradation for only half a month assuming that the chemical was applied uniformly throughout a month.

The values for the factor, APLOSS, were assumed to be 0.15, 0.25 and 0.25 for aerial, ground, and other modes of application respectively based upon recovery data (Akesson et al., 1972). Fortunately, the model is not very sensitive to the value of APLOSS as will be shown later. The monthly use of the chemical was stored in separate arrays according to its mode of application, corrected with appropriate values of APLOSS, and then subjected to the degradation calculation. The first order rate constant, k, for the degradation of parathion in soil was calculated from its half life, t, as k = 0.69/t. The values for t vary considerably with soil type, application rate, and experimental conditions under which they were evaluated. An extensive literature survey revealed that under the experimental soil conditions and application rates similar to those in the San Joaquin Valley, the values for t (Table 1) were scattered approximately according to a log-normal distribution, with mean of log t being -0.291 and standard deviation (SD), 0.245. Three values of k, designated "best estimated", "pessimistic", and "optimistic", were evaluated from mean log t and mean log t plus and minus one SD respectively (Table 2). The model was programmed on a Burroughs 6700 computer and the input parathion use data were read directly from the Data Bank tapes.

SIMULATION RESULTS

The simulation model was executed for the parathion use data of Fresno County from 1968 through 1972. Monthly use, cumulative use, and the three residue levels corresponding to the three degradation rate constants were printed out as tables and figures. The computer print-out for sections in a township range in Fresno County is not shown here, but is available on request. The simulation results were displayed in a six by six grid corresponding to a six mile by six mile township range with each cell of the grid representing a one $mile^2$ section. Figure 3 is a typical

reproduction of the print out of a cell. This particular one is
for parathion used in August, 1968, in section #7, range 15 E,
township 17 S in Fresno County. Total use of parathion for the
month was 38 lb, all applied aerially. After correction for APLOSS,
21 lb reached the soil as calculated using equation 1, the cumulative
use of the year until August is 28 lb, indicating no chemical was
used before August. At the end of August, the residue in soil was
estimated by the model to be 3.6493, 8.9757, or 11.1817 lb
respectively, if 3.5 month^{-1}, 1.7 month^{-1}, or 1.14 month^{-1} was used
as the degradation rate constant in equation 2. At present, this
type of print-out can be produced for each section of every township range in all the counties of San Joaquin Valley for each month
from January, 1968 through December, 1972.

Similar simulation can also be done on a county as a unit.
Figures 4a, b, and c show the results for Fresno County as a whole
for the five year period from 1968 through 1972. The initial
condition for the residue levels were obtained using three k values
to represent "best" (Figure 4a), "optimistic" (Figure 4b), and
"pessimistic" (Figure 4c) estimations. In all cases, the residues
in soil fluctuate closely along with the monthly use pattern,
reflecting rapid degradation of the chemical, and, at the end of
1972, they all approach zero, indicating no accumulation of
parathion in the soil.

Behavior of the first-order degradation model dictates that
the higher the application rate or the residue level, the faster
the degradation rate; and once the amount of monthly use is reduced
or ceases, the residue dissipates in a definite period of continuous degradation. This signifies that the model output is not
sensitive to the initial condition, but rather to the application
pattern. Since parathion use in Fresno County is concentrated
in spring and fall, allowing the residue level to approach zero
toward the end of winter, the zero initial condition assumption
is reasonable if the simulation is started from the beginning of
an application year.

The model, as mentioned previously, is not sensitive to the
value of APLOSS. The stability of the model is evident from
the simulation results using APLOSS values ranging from mean plus
50% to mean minus 50% (Figure 5). The simulation results do not
vary significantly due to the variation in the values of APLOSS,
and, more importantly, their general trends remain unchanged.

                              DISCUSSION

According to the results of calculation (Figure 4): 1) the
increase in parathion use as an aftermath of DDT banning seems

TABLE 1

The Half-life of Parathion in Soils Selected for Simulation

Reference	Half-life[a] (month)	Soil Condition	Application Rate
Harris (1970)	0.23	Sandy loam soil	normal
Iwata et al. (1973)	0.33	Mocho silt loam - 40% saturation	20 ppm
	0.27	Linne clay - 43% saturation	20 ppm
	0.50	Madera sandy loam - 41% saturation	20 ppm
	0.50	Laveen loamy sand - 39% saturation	20 ppm
	0.50	Laveen loamy sand - 43% saturation	20 ppm
	2.00	Santa Lucia silt loam - 33-53% saturation	20 ppm
	0.80	Lucia silt loam - 43-63% saturation	20 ppm
	1.60	Windy loam - 47-73% saturation	20 ppm
	0.43	Madera sandy loam - 39% saturation	200 ppm
	0.43	Madera sandy loam - 42% saturation	200 ppm
Kearney et al. (1969)	0.13	Normal agricultural soil	normal
Knutsen et al. (1971)	1.00	Silty clay loam	1 lb/acre

Lichtenstein & Schultz (1969)	0.60	Carrington silt loam	5 lb/acre
Sacher et al. (1972)	0.90	Ray silt loam - 20% on vermicullite	1 lb/acre
	0.52	Ray silt loam - 10% on vermicullite	1 lb/acre
	0.47	Ray silt loam - 5% on vermicullite	1 lb/acre
	0.20	Ray silt loam - kept at 24°C - daily watering	1 lb/acre
	0.70	Ray silt loam - 10% on attapulgite - seasonal average	6 ppm

a Half-lives were calculated from reported data.

TABLE 2

Rate Constants (k) for the First Order
Degradation of Parathion in Soil as Determined
From the Log-normal Distribution of Half-lives (t).

Estimate	log t	(month)	K = 0.69/t (month^{-1})
Best estimate	mean = -0.291	0.512	1.35
Optimistic	mean - SD[a] = -0.568	0.260	2.66
Pessimistic	mean + SD = 0.003	1.009	0.684

[a] SD = standard deviation of log t.

Section # 7

Air	= 2.8000E 01
Ground	= 0.
Other	= 0.
Total	= 2.8000E 01
Net	= 2.1000E 01
Acc.	= 2.8000E 01

Rate Constant = 3.50
Res   = 3.6493E 00

Rate Constant = 1.70
Res   = 8.9757E 00

Rate Constant = 1.15
Res   = 1.1817E 01

Figure 3. Parathion use data and estimated residues in soil in
a section.

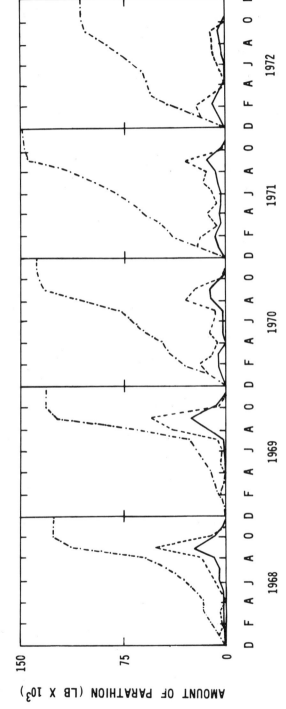

Figure 4a. The simulation output showing the annual cummulative (-•-•-) and monthly use (---) of parathion, and the average residues in soil (———) of Fresno County. Residue estimated using the "best estimated" k value, 1.35 month^{-1}.

210

D. P. H. HSIEH ET AL.

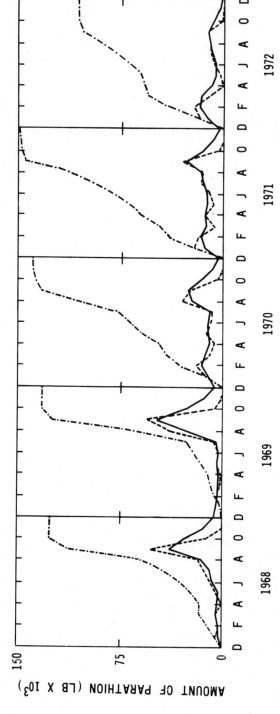

Figure 4b. The simulation output showing the annual cummulative (-•-•-) and monthly use (----) of parathion, and the average residues in soil (———) of Fresno County. Residue estimated using the "optimistic" k value, 2.66 month-1.

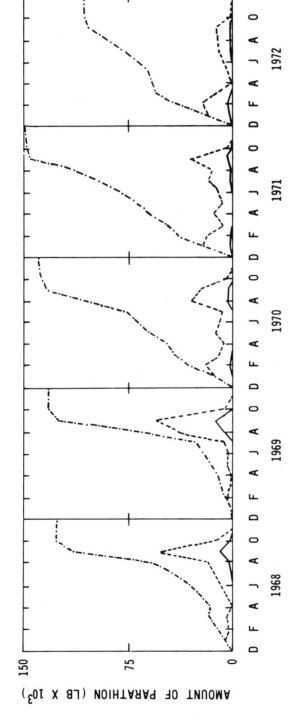

Figure 4c. The simulation output showing the annual cummulative (-•-•-) and monthly use (----) of parathion, and the average residues in soil (——) of Fresno County. Residue estimated using the "pessimistic" k value, 0.684 month⁻¹.

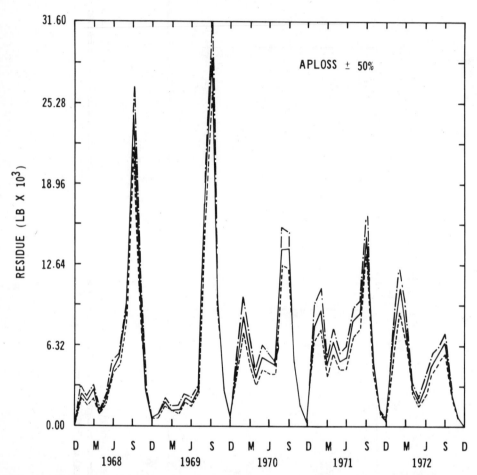

Figure 5.   Sensitivity of the simulated residue level with respect
to APLOSS values.   The "best estimated" k value used in this trial
along with 3 APLOSS values: APLOSS (———), APLOSS + 50% (-•-•-),
APLOSS - 50% (----).

to be over, and the annual use of parathion is probably going down from a peak in 1971 or 1972; 2) at the present schedule and rate of parathion use, the residue level in soil oscillates with respect to time, approaching a negligible low value at the end of the slow season; and 3) as long as there is no significant increase in use, parathion is not accumulating in the soil; and the major residue problems are short term and acute in nature.

As mentioned previously, the present model is an exceedingly simplified simulation. It does not take into consideration the effects of various factors on the degradation of a chemical in soil. The deterministic values have been assigned to application loss, and rate constants based upon incomplete experimental data of some special cases. A section, or even the whole county, is regarded as a homogeneous system, which it is not. These are the areas of shortcomings in the structure of the model which need continuing, stepwise refinement to narrow the discrepancy between what happens in the real world and what is predicted by a simulation model.

Despite the simplicity of the model, until adequate experimental information becomes available for refinement, it serves the purpose of mathematical simulation as to translate qualitative terms into quantitative terms to make the analyses more objective. The model also provides an educated guess as to the residue situation of parathion under the current agricultural use pattern in the San Joaquin Valley.

SUMMARY

Recent restrictions on use of DDT resulted in a considerable increase in the use of parathion in the San Joaquin Valley. A simple mathematical model was programmed into a Burroughs 6700 computer to estimate the rate of accumulation of parathion residues in soil over the period of 1968 through 1972. Actual use data, the input to the model, are made available from a Data Bank developed and maintained in this center. The model takes the form of a general mass balance equation:

Accumulation = Application - (application loss + first order degradation)

A one mile x one mile section of a county is used as the system of analysis, and month is the time unit. Three values for application loss were used for aerial, ground, and other modes of application. Three degradation rate constants were used to correspond to optimistic, pessimistic, and best estimates. The model outputs indicate that parathion residues are not accumulating in the soil, and that the problems associated with the current agricultural use of parathion are acute and short term in nature.

ACKNOWLEDGEMENT

The authors acknowledge the valuable contributions of Robert W. Cooper, deceased. They also thank David E. Wedge for assistance in computer programming. This study was supported by NSF Grant #GB-33723.

REFERENCES

Akesson, N.B., W.E. Yates, and P. Christensen 1972. Aerial dispersion of pesticide chemicals of known emmissions, particle size and weather conditions Paper presented at the American Chemical Society Meeting, Boston, MA.

Edwards, C.A. 1966. Insecticide residues in soils. Res. Rev. 13:83-132.

Hamaker, J.W. 1972. Decomposition: quantitative aspects. In Organic Chemicals in the Soil Environment G.A.I. Goring and J.W. Hamaker editors. Marcel Dekker, Inc., N.Y.

Harris, C.R. 1970. Persistence and behavior of soil insecticides. In Pesticide in the Soil. Michigan State University, East Lansing, MI.

Hsieh, D.P.H., T.E. Archer, D.M. Munnecke, and F.E. McGowan. 1972. Decontamination of noncumbustible agricultural pesticide containers by removal of emulsifiable parathion. Environ. Sci. Technol. 6:826-829.

Iwata, Y., W.E. Westlake, and F.A. Gunther. 1973. Persistence of parathion in six California soils under laboratory conditions. Arch. Environ. Contam. Toxicol. 1:84-96.

Kearney, P.C., R.G. Nash, and A.R. Isensee. 1969. Persistence of pesticide residues in soils. In Chemical Fallout: Current-Research on Persistene Pesticides. M.W. Miller and G.C. Berg editors. Charles C. Thomas, Ill.

Kilgore, W.W., N. Marei, and W. Winterlin. 1972. Parathion in plant tissues: new consideration. In Degradation of Synthetic Organic Molecules in the Biosphere. National Academy of Sciences.

Knutsen, H., A.M. Kadoum, T.L. Hopkins, G.G. Swoyer, and T.L. Harvey.
    1971.   Insecticide usage and residues in newly developed
    Great Plains irrigation district.   Pestic. Monit. J. 5:17-27.

Lichtenstein, E.P., and K.R. Schultz.   1964.   The effects of
    moisture and microorganisms on the persistence and metabolism
    of some organophosporous insecticides in soils, with special
    emphasis on parathion.   J. Econ. Entomol. 57:618-627.

Sacher, R.M., G.F. Ludvik, and J.M. Deming.   1972.   Bioactivity and
    persistence of some parathion formulations in soils.   J. Econ.
    Entomol. 65:329-332.

Spector, W.S.   1956-59.   Handbook of Toxicology.   W.B. Saunders
    Co., Philadelphia PA.

Walker, A.   1973.   Use of a simulation model to predict herbicide
    persistency in the field.   Proc. Eur. Weed Res. Coun. Symp.
    Herbicides-Soil 240-250.

Wolfe, H.R., D.C. Staiff, J.G. Armstrong, and S.W. Comer.   1973.
    Persistence of parathion in soil.   Bull. Environ. Contam.
    Toxicol. 10:1-9.

PARTITIONING AND UPTAKE OF PESTICIDES IN BIOLOGICAL SYSTEMS

Eugene E. Kenaga

Health and Environmental Research Department

The Dow Chemical Company, Midland, Michigan 48640

INTRODUCTION

Biological systems can be interpreted to mean the mixture of things which are necessary for life, whether inside or outside of a single organism or community of organisms. Some of the major necessities for life include water, oxygen, organic food, minerals, energy generated by the sun, flow systems, and a media to live in. All chemicals making up such necessities exist in gas, liquid, and solid phases in various ratios which are temperature, pressure, and solubility dependent. These phases which are always striving for a balance in complex biological systems can be pictured as shown below:

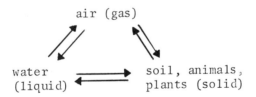

air (gas)

water (liquid) ⟶⟵ soil, animals, plants (solid)

Within biological systems equilibrium is rarely if ever reached since organisms in order to live must use energy to replace cells, grow in size or numbers, respire, and excrete and therefore demand a constant flow in respiration nutritional and excretion systems. With plants and animals, the major liquid flow systems are supplied by water externally and internally by water in the form of blood in animals and sap in plants. Uptake and distribution of chemicals in biological systems, whether within or between organisms and their

external environments, involves several phases which can be pictured
simplistically as:

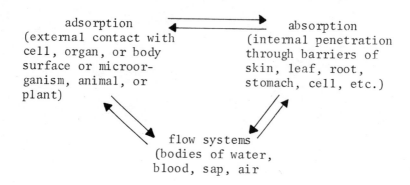

adsorption   absorption
(external contact with   (internal penetration
cell, organ, or body   through barriers of
surface or microor-   skin, leaf, root,
ganism, animal, or   stomach, cell, etc.)
plant)

flow systems
(bodies of water,
blood, sap, air

    The tendency of a compound to adsorb onto a given surface is
related to its inherent chemical and physical properties and is
reproducible under standardized test conditions. Chromatographic
analytical methods are used by chemists in separating and identify-
ing compounds from a mixture of chemicals. In a flow of air (gas)
or water (liquid) a given compound is retained at a given place on
an adsorbent surface in the flow system after a given period of
time. Similarly the separation of sediments and dissolved materials
in water in nature is analogous to a giant chromatographic system
in the way in which it separates compounds by adsorption on various
natural surfaces. The ratios of separation of a given chemical
between liquids, solids, and/or gases have been described as partition or distribution coefficients or as adsorption coefficients
between solids and liquids, or solids and gases. The properties of
chemicals which are key indicators for estimating typical distribution coefficients are solubility (especially in water and fat
solvents), vapor pressure, and the interfacing surface energy forces,
whether mutually attractive or repellent (anionic, cationic, pH,
polarity, and other electrical forces).

    Estimation of distribution coefficients of simple systems (i.e.,
1 ppm 2,4-D partitioning in 1:1 ratios of n-octanol and water by
volume in the liquid phase, is useful for isolating affects of
individual components of an ecosystem. However, aside from changing
ratios of the three chemicals which may occur in real life aquatic
situations, an infinite number of variables such as weather, variable flow speeds, kinds of water containing immeasurable kinds of
solid and soluble impurities, etc. must be factored into the distribution equation.

    For the above reasons, the subject of this paper which is the
distribution and uptake of pesticides in biological systems is a
vast and complex problem. The author has chosen to emphasize

aqueous systems, because water is the main control to the distribution process.

AQUATIC FLOW SYSTEMS

Bodies of Water

Water is the principle transport fluid on the earth's surface
and covers 70 percent of the earth's surface. All life must deal
with it. Consequently, its physical and chemical properties are
important. Water has unusual properties compared to organic solvents such as being at its highest density at 4°C, rather than at
its freezing point. It also has an unusually high boiling point,
viscosity, and surface tension, compared to its molecular weight.
Water is never chemically pure in nature. Crosby (1972) has pointed
out some of the impurities in lake and ocean water (Table 1).

Natural water contains dissolved oxygen, carbon dioxide, inorganic salts (3.5% in sea water) and trace elements. It often
contains ammonia, organic and inorganic dust held in suspension
and dissolved organic matter (such as ketones, alcohols, fatty
acids, aldehydes, humic acids, common amino acids), and microorganisms as well as other natural and man-made chemicals. Water
can dissolve many chemicals from a minute (ppb) amount to more than
equal volumes of water, tending to dissolve polar compounds much
more readily than non-polar compounds.

Water is never still even in quiet ponds since thermal currents
generated by daily, diurnal, and seasonal changes cause water turn-
over, as well as those caused by wind. The water flow of fast
running rivers going over cascades or falls represents the maximum
speed of movement of water, aeration of water and opportunity for
volatility of pesticides from water.

Blood and the Distribution of Food and Water in Animals

The blood of animal systems can be likened to a river of buffered water and contains many essential molecules or waste metabolites
or introduced in non-essential foreign molecules. Many of the nutrients carried by the blood are gradually accumulated in much higher
concentrations in specialized metabolic and deposition areas of the
body.

Blood contains all of the essential substances necessary for
the existence and development of the cell and body. In higher animals, blood consists of a fluid, the plasma, in which are suspended
corpuscles of various kinds, adapted for special purposes. The

TABLE 1

Impurities in Fresh and Ocean Water (Crosby, 1972)

	Lake	Ocean
Typical surface temperature, °C	20	12 (San Francisco)
		18 (Los Angeles)
pH	5.0-9.0	7.6-8.4
% T_{300}, 1 ft depth[a]	85	85
$[O_2]^{15°}$, m$\underline{M}$	0.30	0.26
$[Cl^-]$, m$\underline{M}$	0.16	550
$[Br^-]$, m$\underline{M}$	0.002	0.8
$[Cu^{++}]$, m$\underline{M}$	Variable	0.0002
Organic C, ppm	10	2-7

[a] Ultraviolet transmission at 300 nm.

composition of the plasma is roughly as follows: water, 90%; proteins (fibrinogen, globulins, albumin), 9%; salts (mainly sodium chloride, potassium chloride, and sodium bicarbonate), 0.9%; sugar, urea, uric acid, creatin, and other trace nutrients, enzymes, hormones, and excretions in addition to phospholipids and other lipids. A primary use of sodium chloride in blood is as a protein solvent which is necessary for the transport of amino acids since some amino acids do not dissolve sufficiently in saltless water. Protein can form the basis of living material only if it is in solution and carried to the cell. The primary purpose of the red blood cell of which there are normally several million per mm^3 of blood is to carry oxygen. Red blood cells contain haemoglobin which is an iron-containing conjugated proteinaceous respiratory pigment superficially related to chlorophyll. White blood cells in vertebrates may normally number 5,000 to 10,000 per mm^3 and serve to ingest various pesticides in the blood including microorganisms. The blood clotting platelets may number 500,000 per mm^3 of blood.

The pH of normal plasma is 7.4. Changes in pH are small when acids or alkalines are added because of the great buffering capacity of the blood. This is achieved principally by the interaction of sodium bicarbonate in the plasma with haemoglobin in the corpuscles.

The blood in a mammal has two main routes of circulation. One route is relatively rapid, that of the renal and portal circulation. The capillary route which furnishes blood to most organs is relatively slow, but may be furnished blood through different sized capillary systems, which in turn have different rates of flow. These various flow rates furnish different diffusion rates for gases and solvents from blood, resulting in different uptake and distribution rates for various chemicals, whether natural or unnatural, in the tissues and organs. Partitioning of introduced chemicals from the blood to tissues would come to equilibrium rapidly in tissues receiving rapid blood flow rates. Thus, in an acute exposure an introduced chemical in a rapid blood flow route would quickly attain maximum concentrations and lose concentrations rapidly upon withdrawal of the chemical. Introduced chemicals in a slow system would show lag periods of time in attaining maximum concentration or clearance of the chemical from such tissues. It is thus nearly impossible to come to a steady state equilibrium of concentration of an introduced compound especially if it is easily degraded in, or diffused from, a specific organ.

The cardiac output of mammals was tabulated by Spector (1956) and as shown in Table 2 varies greatly in total flow between species but only by a two-fold factor on the basis of body weight.

The rapidity of (the shortest complete cycle of) circulation of blood in mammals is illustrated by that of humans, 7-9 seconds; dog, 10-11 seconds; and sheep, 5-8 seconds. Circulation to various organs of the body to another, whether mammal or bird, can be from one second to one minute (Spector, 1956).

Davis (1970) found the mean value of the blood circulation time to be 64 seconds in rainbow trout in water at 10°C.

Some other body conditions which are affected by blood, and which blood affects are body temperature, and pH of the cells, tissues and organs of animals. The body temperature of 165 species of mammals ranged from 25.0-40.4°C (rectal) and the body temperatures of 160 species of birds ranged slightly higher from 37.8-44°C (cloacal) (adapted from Spector, 1956).

The temperatures are related to metabolism of body food, and pesticides, rates of hydrolyses, enzymatic reactions, vapor pressure, etc.; all being generally higher as the temperature increases. Fish, amphibia, reptiles, and invertebrates are cold-blooded. Generally closely assuming the temperature of their surroundings.

TABLE 2

Cardiac Output of Vertebrates (Adapted from Spector, 1956)

Species	Body Wt kg	Blood Flow 1/min Basal	Blood Flow 1/min kg Body Wt (Average)	Blood Capacity ml/kg[a] Body Wt	Blood Capacity Total 1[b]
Man	66-72	4-5.1	0.064	69-80	5
Dog	6-16	1.12-2.21	0.15	92.6	0.9
Horse	283	18.8	0.066	72	20
Rabbit	4.13	0.53	0.128	50	0.21
Goat	23.7	3.1	0.13	60-90	1.8

[a] Altman (1961).

[b] Calculated on basis of body weight.

Spector (1956) has tabulated the pH of a number of cells, tissues, and organs of various phylla of animals and plants. pH values in animal organisms vary from 3.5 (digestive vacuoles of protozoa) to 8.5 (cytoplasm of fertilized egg of a frog). However, most values in the 14 phylla fell between pH 5 and 7.6. In mammals the pH values for the six species cited ranged from pH 6.0 to 7.6. Even with single organs such as the liver pH values in peripheral cells were slightly alkaline while those in the central cells were slightly acidic. The digestive system also begins usually with an acidic stomach and progresses to an alkaline condition in the large intestine.

Water is the principal component of many species of organisms. Insensee et al. (1973) measured the water content of algae, snails, daphnids, and fish as 65.7, 72.4, 84.9, and 75.6%, respectively.

Water is the largest component of most tissues and organs of animals. From the compilation of Spector (1956) and as summarized in Table 3, it can be seen that among the major metabolic tissues such as nerve, lung, heart, gonads, bone, muscle, liver, and kidney, the water content ranges from 52-89%; the protein content from 8-25% and the lipid content from 0.2-20%. Contents of protein may be much

TABLE 3

The Approximate Ranges of Water, Protein, and Lipid in
Animal Tissues (Adapted from Spector, 1956)

Tissue	% Water	Fresh Tissue Approximate Content	
		% Protein	% Lipid
Nerve	60-85	8-12	2-20
Lung	77-85	15-17	
Heart	71-85	16-18	0.2-19.5
Liver	67-75	13-17	2.0-6.5F
Muscle	52-84	14-25	2-32
Kidney	74-80	15-18	1.1-5.3
Gonads	73-89	--	1.1-12.3
Whole Teeth	4-14	18	
Bone	30-81F	--	5-13F
Skin	60-75	25-34	0.3-25
Hair	<4-28	81-91	--

F = Fat tissues

higher is deposition areas such as in skin, and of fat in fat depots.
Unlike plants, animals have little carbohydrate in the form of starch
and sugar in body tissues.

Very little is known concerning the accurate predictability of
partitioning of compounds in various organs of animals. The reason
for this is apparent because of the number of variables involved and
the difficulty in determining steady state conditions and rates of
partitioning through tissues which are complex in function and composition. The rat is a mammal used extensively as a toxicological
index organism for man. Data gathered by Torkelson et al. (1966)
(Table 4) for the distribution of water and lipids, and blood flow
rates of rats can be used to illustrate certain generalities, and
demonstrates the variability within a single species. Since distribution rates vary with exercise and changing metabolism, the data
may vary accordingly.

TABLE 4

Percent Water and Lipid Wet Tissues of Rats and Rate of
Blood Flow Through Its Organs

Tissue Organ	% in Wet Tissues		Blood Flow Through Organ[b]	
	Lipid	Water[a]	ml/min/organ	ml/min/g
Blood	0.5^a	84.0	--	--
Serum	1.13^b			
Heart	4.25^a	85.4	1.15-1.34	1.81
	$0.4-3.06^b$			
Lung	$3.14-4.4^b$		1.20	0.67
Testes	2.28^a	91.6	--	0.14-0.21
	$1.1-3.0^b$			
Liver	4.17^a	71.1	3.1	0.627
	$4.3-6.3^b$			
Kidney	3.80^a	83.7	7.76-8.20	4.2
	5.2^b			
Muscle	3.8^b			
Fat	76.5^a	20.5	0.088^c	
Gut	--	--	8.54-8.59	0.817
Brain	$7.1-9^b$		0.56-0.63	
Skin	8.13^b		3.10-4.0	0.126
Whole body	18^b			

[a] Original determinations.

[b] Cited data.

[c] Epididymal fat pad.   Torkelson et al. (1966).

    1. The main component in the body is water, regardless of organ or tissue, with minor exceptions.

    2. Lipids are another large component of animal tissues. While variable in content they are found in every major organ and are

present in a concentration of about 0.5% in the blood. Fat depots
may contain 76.5% lipid (and 20.5% water). Skin and brain are
also high in lipid content.

   3. Blood circulation is not equally rapid through each organ,
varying from low levels in fat, skin, and testes to high levels in
kidney, heart, and gut and with changing activity and metabolism.

The rate of digestion of food and passage through the digestive
tract is related to the speed and degree of uptake of food and pesticides in food by the blood. The speed and assimilation of these
chemicals depends on many things which will not be discussed here
since time and space limits a full evaluation of this important subject matter. Suffice for now to point out that while it may take
several days to digest food in some animals species it may only be
a matter of minutes in others and thus assimilation of pesticides
must be rapid. In the case of an insoluble pesticide very little
absorption may occur.

Domestic fowl may pass grain through the digestive tract in
2.5-12 hours depending on the type and amount of food and the physiological state of the individual (Sturkie, 1965). Ordinarily a bird
digests animal food more rapidly. For example, a magpie (*Pica*) may
digest a mouse in three hours (Hewitt, 1948). Seeds of berries
eaten by young waxwings appear in their feces in as little as 16
minutes and thrushs will defecate seeds of elderberry within 30
minutes.

Plant Sap and the Distribution of Food and Water in Plants

In woody plants, water enters mostly through the roots carrying
minerals and is drawn by osmotic pressure through the xylem to the
leaves where sugars, proteins, and soluble foods are formed and then
move to growing points and downwards in the phloem (and xylem) even
to the roots. Various substances such as water, food, and waste
products accumulate in the stems and roots. These substances include
carbohydrates, chiefly starch, fat, and amino acids, and many other
substances such as musciliage, resins, essential oils, latex, lignin,
and cellulose. Plants do not parallel animals as far as excretion
is concerned.

Plants contain mostly water. Over 50 percent of the total fresh
weight of a tree consists of water but the content varies widely in
different parts of a tree, and also with species, age, site, season
and even the time of day. The leaves of growing plants may contain
as much as 90% water (Kramer and Kozlowski, 1960).

Water is one of the raw materials in photosynthesis in plants
and is a vehicle for transport of mineral salts from soil to places

of utilization within the plant and/or organic solutes from regions
of synthesis to those of assimilation or storage. For example, the
total water absorbed by forage plant roots may be 300-800 times the
dry weight accumulated by the plant and is transpired mainly as
water vapor. Water in cells or sap of plants is highly buffered,
the salts of phosphoric, malic, and citric acid being among the most
important substances controlling or regulating pH.

While plants do not have a complete circulation system, they
do move sap very rapidly under some circumstances, particularly at
midday. In diffuse porous species such as maples and beech the
velocity ranges from 1 to 6 meters per hour, while in ring-porous
species such as oak it varies from 25 to 60 meters per hour in the
xylem (Kramer and Kozlowski, 1960). Xylem translocation (of minerals in sap) occurs upwards through dead elements and is rapid, while
translocation of minerals and organic substances through phloem is
through living sieve elements and is slower.

The classification of trees has been based on these storage
characteristics at mid-winter. Some trees such as pine and linden
may contain as much as 6 to 7 percent fat. Nutmeats and seeds of
pines may contain over 70% oil. Roots of trees contain little or
no fat. Leaves rarely contain over 5% and often have much less.
Some trees have low fat and high starch contents. Carbohydrates
constitute about 75% of the dry weight of trees and 1-20% of the
dry weight may be starch and sugar.

A group of 80 fresh uncooked foods of plant origin were listed
by Spector (1956) according to composition. This list was divided
into food groups on the basis of relatively similar water, protein,
fat, and carbohydrate proportions as shown in Table 5. Freshly
harvested grain and legume seeds and nuts were the only foods to
contain less than 40% water, more than 20% protein, and more than
35% carbohydrates. Fatty nuts and fatty juicy fruits contained
26-73% fat. Vegetables and 'juicy" fruits contained mostly water,
with carbohydrates being second highest in percent.

Plant cells and tissues of the terrestrial species tabulated
(Spector, 1956) were much more acidic than those of animal species
and varied in pH from <3.4 to 6.8. Most values were between pH 4.0
and 5.9. These cited values do not include variations of pH in or
on plants and animals involving specific surface areas or cells
which may show greater variation than those cited. The cell sap
of leaves of 45 species of plants measured for pH were acidic and
ranged mostly between 5-6.9. A few species were extremely acidic
such as *Begonia*, 0.9-1.4; *Oxalis*, 1.7-2.1; and rhubarb, 3.6.

TABLE 5

Water, Protein, Fat, and Carbohydrate Distribution of Fresh
Uncooked Food (Adapted from Spector, 1956)

Food Group	Approximate Range of Composition (%)			
	Water	Protein	Fat	Carbohydrate
Grain and legume seeds	11-12.6	7.5-23.8	1-1.8	59.4-78.8
Fatty nuts	2.6-5.3	3.4-26.9	34.5-73.0	11-23.6
Vegetable--green foliage, stems, roots	74-94	0.1-7.5	0.2-1.4	2.9-31.2
"Juicy" fruit (pome, citrus, berries, etc.)	74-94	0.3-1.4	0.1-1.4	4.0-14.9
Fatty, juicy fruits (coconut and avacado)	45-67	1.7-3.4	26.4-34.7	5.1-14.0

Comparison of Natural Water, Blood, and Plant Sap and the
Distribution of Food and Water in Plants and Animals

Comparisons of natural water, blood, and sap show certain
general similarities in that they all consist mainly of water with
"impurities," have extremes of flow rates from slow to relatively
fast, are usually quite well buffered, contain inorganic salts, and
organic carbons. The pH of natural water fluctuates widely from
acid to alkaline. The pH of blood is nearly neutral while the pH
of sap is generally well below pH 6. Plants and animals fed nutrients by sap and blood respectively, both may accumulate lipids or
proteins in small to large amounts. The carbohydrate content of
plants is unusally high while that of animals is low.

DISTRIBUTION OF PESTICIDES IN AIR, WATER, SOIL, AND BIOTA

Pesticide Volatility From Water

Volatility from water is important because, as is noted in the previous section, it might also apply to volatility from blood or plant sap, as well as from wet soil or green (wet) foliage and is a major cause of loss of many pesticides from such media or surfaces.

Dilling et al. (1973) studied the evaporation rate of chlorinated hydrocarbons at 1 ppm (well below their solubility in water) using mass spectrometry for analyses. Methylene chloride, chloroform, 1,1,1-trichloroethane, trichloroethylene, and tetrachloroethylene, five compounds which were studied in detail, had evaporated about 50% in less than 45 minutes, and about 90% in less than 100 minutes from water when stirred (200 rpm) at 25°C in an open container. Addition of various water contaminants (bentonite clay, 375-500 ppm; limestone, 500 ppm; sand, 500 ppm; sodium chloride, 30,000 ppm; peat moss, 500 ppm; and kerosene, a thin layer) to water had some, but relatively little effect, on the rates of evaporation or disappearance of these chlorinated solvents from water. The selected contaminants are representative of particles found in various bodies of water which might act as sorbents and thus reduce the availability of the solvent for volatilization. These solvents are all more volatile than water at 25°C (23.79 mm Hg) except tetrachloroethylene which is similar. Thus, in the case of these solvents, the vapor pressure appears to negate the importance of the added solid sorbent and solvent effects.

Using a 2.2 mph wind (from a fan) across the surface of the water for 20 minutes, the evaporation of these solvents was about 17% greater than in still air. At 1-2°C there was a 28% decrease in the amount of these solvents evaporated after 30 minutes than at 25°C.

Salts are essentially non-volatile. Hardy (1965) found that picloram potassium salt, at 1 ppm picloram in water did not volatilize significantly from aquatic ecosystems, although considerable water loss from the test aquaria occurred over a period of ten weeks. Smith et al. (1972) studied the distribution of 5 ppm dalapon sodium salt in a aquatic ecosystem with little or no loss of concentration of dalapon by water evaporation.

Kenaga (1972) noted the decreased percentage of volatilization from water when the water solubility of compounds was exceeded.

Goring (1967) calculated the ratio of weights of chemical in equilibrium in equal volumes of water and air at approximately the same temperature. The chemicals were assumed to follow the gas laws approximately, and concentration in air at the listed vapor pressure were assumed to be in equilibrium with the concentration in water at maximum solubility in water. The water to air ratios of some representative compounds were carbon disulfide, 1.8; methyl bromide, 4.1; DDT, 303; diazinon, 16,000; parathion, 30,800; diuron, 990,000; atrazine, 18,430,000. From these data it is apparent that fumigants and DDT are relatively quite volatile from water, even though greatly different in molecular weight and vapor pressure, while atrazine is not very volatile.

The volatility of organic compounds from water generally increases with increasing amounts of water evaporation; increasing vapor pressure of the compound; and decreasing water solubility. Volatility from water of pesticides such as DDT, which are relatively insoluble in water, is greatly reduced as a result of sorption by water and soil constituents, particularly soil organics, lipid surfaces, or animal organisms.

Goring (1967) and Hamaker (1972) have used water solubility and vapor pressure to calculate water to air ratios (partition coefficients) of a number of pesticides. These data are useful for comparing the tendency of compounds to volatilize from water in the absence of solids dispersed in water. The amount which might volatilize from a multi-component system is more difficult to calculate since sorption by solids must be taken into account as well as the ratio of the volumes of each component. Goring (1972) has made such calculation for soil fumigants.

Mackay and Wolkoff (1973) have developed an equation for calculation of the rate of evaporation of low solubility contaminants for water bodies to atmosphere. The rate of evaporation of hydrocarbons and chlorinated hydrocarbons can be high even for compounds of low vapor pressure such as DDT, PCBs, and n-octane.

The vapor pressure of a compound can be measured by standardized techniques and expressed as a precise figure at a given (room) temperature at sea level pressure (760 mm Hg). These data are useful where molecules are volatilizing from a liquid or solid mass of the same compound. However, when a pesticide is applied at a uniform dosage rate to heterogeneous surfaces such as natural water, soil, foliage, wood, or glass, volatility is variable because of variability in sorption. A compound such as chlorpyrifos which has a low vapor pressure can be very persistant on dry wooden surfaces, and in dry organic soil and relatively non-persistent on green foliage surfaces as judged by bioassay tests (Kenaga et al., 1965). A compound such as tricyclohexyltin hydroxide which in the dry state has a vapor pressure so low that it is essentially unmeasurable can be shown to kill spider-mites infesting a bean plant by transfer of its vapors from the inside of a treated glass jar covering, but

not touching the plant (Whitney and Kenaga, 1966). By use of
radioactive [119]Sn, tricyclohexyltin hydroxide was later shown to
codistill with water, thus explaining the vapor transfer to the
mite-infested foliage indicated by the bioassay (Smith et al., 1970).

From these tests and others in soil, it is shown that the increased volatility of relatively non-volatile materials is due in
part to the presence of water from which the solubilized pesticide
(even though low in water solubility) may volatilize easier than
from a dry surface.

Pesticide Sorption on Soil or Carbon

The sorption of pesticides in soil as related to soil characteristics has been thoroughly reviewed by numerous authors (Goring,
1967; Bailey and White, 1970; Weber, 1972; Stevenson, 1972; Goring
and Hamaker, 1972; and others). The reason for brief discussion
here is to point to certain similarities in sorption potentialities
between certain segments of the soil and plants and animals since
they are competitive sites for sorption of natural chemicals and
pesticides, especially in aquatic ecosystems.

Bailey and White (1970) have reviewed the literature concerning
factors influencing the adsorption, desorption, and movement of
pesticides in soil. Many of the same factors are basic and are useful in understanding the adsorption of pesticides to plants, animals,
and microorganisms. Among the physico-chemical factors affecting
sorption between the adsorbate and adsorbant are size, shape, con-
figuration, and total area of the molecule or particle, acidity,
or basicity of the molecule or surfaces (pKa or pKb), water solubility, charge distribution, polarity, polarizability, reactivity
in order to undergo hydrogen bonding, chelation, polymerization,
complexings, conjugation, temperature, concentration or dilution,
and weathering.

Organic matter has been shown to be the most important sorptive
surface in soil for all but a few highly ionic pesticides, chiefly
herbicides. Even though mineral particles predominate they are often
coated with organic substances. Most soils contain some organic
matter ranging mostly between 0.1% to 7%. An analysis of the functional groups of two organic soils (Ahrlichs, 1972) showed the presence of carboxy, carbonyl, alcoholic hydroxyl, and phenolic hydroxyl
groups. The general composition of soil organic materials is shown
in Table 6. Thus, the heterogeneous multisorptive nature of soil
organic material should not be surprising.

Wershaw et al. (1969) suggested that sodium humate, a major
natural organic constituent of soil, lowers the surface tension of
water much like sodium lauryl sulfate (a strong anionic surfactant)

TABLE 6

Composition of Soil Organic Materials (Weber, 1972)

Type of Soil Organic Matter	% of Carbohydrates	% of Soil Organic Matter
Carbohydrates		5-15
Hexanes		
Pentoses		
Oligosaccharides		
Polysaccharides (celluloses)	10-20	
Amino sugars		
Sugar alcohols		
Sugar acids		
Methylated sugars	<1	
Free carbohydrates	<1	
Unidentified	70-80	
Humic materials		85-90
Humic acids		
Stable free radicals		
Fulvic acid		
Humin		

suggesting that sodium humate also might help solubilize compounds
which are highly water insoluble. Tests with DDT show that a 0.5%
sodium humate concentration increases solubility in water at least
20 times up to 4 ppm. In contrast to sodium humate, which is highly
soluble in water, humic acid is practically insoluble in solutions
of pH less than 7 or 8. Humic acid strongly sorbs 2,4,5-T from
solution. The solubilization of DDT, a nonpolar compound, by sodium
humate, and the sorption of 2,4,5-T, an ionic compound, by humic
acid are two varied types of solubility interactions which might be
expected when any organic pesticide is applied to a natural organic
soil-water system.

Weber and Gould (1966) studied the adsorption of 2,4-D, 2,4,5-T,
silvex, carbaryl, and parathion on active carbon 273 microns in size
for use in pesticide removal from water. The rate of adsorption for
all compounds appeared to be linear as a function of the square root
of time starting from the time the pesticide was brought in contact
with the carbon adsorbant, at least during the first nine hours.
This means that most of the adsorption occurred in the first few
hours. As shown in Table 7, at least 75% of 2,4,5-T and parathion
were sorbed in the first 16 hours of the exposure. The effect of
increase in concentration of the pesticide in water appears to be
correlated with an increase in the rate of adsorption in terms of
mg of pesticide/g of carbon. Adsorption does not appear to follow
completely in accordance with the monomolecular layer assumed for
the Langmuir model for adsorption equilibrium. The weight of 2,4,5-T
for example, adsorbed by carbon (up to 39%) far exceeds that which
would occur on a monomolecular layer basis (Table 8).

The ratio of mg/g 2,4,5-T adsorbed by charcoal to ppm 2,4,5-T
in water at equilibrium time decreases at a rapid rate from 1 to 15
ppm 2,4,5-T in water, representing a decrease in adsorption efficiency as the concentration increases. "While the Langmuir formula
does adequately represent data over a large and relatively high
concentration range it seems likely that more refined experimentation
in a lower concentration range (say 1-100 ppb) would show considerably greater adsorption of the pesticides than that predicted by
the Langmuir constants."

Hamaker and Thomson (1972) have constructed a useful outline
for representative pesticides relating their molecular class (acid,
base, polar, or neutral) and form (anion, cation, free acid, free
base, or un-ionized) with effect of low and high pH on soil adsorption (Table 9) of these pesticides.

Goring (1967) and Hamaker and Thompson (1972) have calculated
soil organic matter to water ratio by use of the ppm of pesticide
adsorbed on soil organic carbon (K_{oc}) per ppm pesticide in water
solution basis. This was done because soil organic (converted from

TABLE 7

Rate of Adsorption of Pesticides on Carbon in Water

Pesticide	Dosage Pesticide ppm in Solution	Dosage Pesticide Total mg	Hours Adsorption Time	Adsorption of Pesticide on Charcoal mg/g	Adsorption of Pesticide on Charcoal % By Wt	Adsorption of Pesticide on Charcoal Total mg	% of Total 2,4,5-T Sorbed Out of Solution on Charcoal
2,4,5-T[a]	2.5	20	1	25	2.5	5	25
			2	37	3.7	7.5	38
			4	50	5.0	10	50[c]
			8				63[c]
			16				75[c]
			32				88[c]
			ca 64				100[c]
Parathion[b]	2.91	23.3	1	32	3.2	6.4	27
			2	48	3.7	9.6	41
			4	64	5.0	12.8	54[c]
			8				68[c]
			16				81[c]
			32				94[c]
			ca 48				100[c]

[a] Based on 9.8 μmoles 2,4,5-T/1 in 8 1 water containing 25 mg of 273 micron Columbia LC carbon/1 at 28°C (extrapolated from Figure 1, Weber and Gould, 1966).

[b] Based on 10 μmoles/1 in 8 1 water containing 25 mg carbon/1 at 25°C (extrapolated from Figure 2, Weber and Gould, 1966).

[c] Extrapolated from first few hours on a linear basis.

TABLE 8

Comparison of Concentrations of 2,4,5-T in Water With
Relative Adsorption on Carbon at Equilibrium[a]

ppm 2,4,5-T in Solution	2,4,5-T on Carbon		Ratio ppm:mg/g
	mg/g	% by wt	
1	155	15.5	155
2.5	260	26.0	104
5	330	33.0	66
10	380	38.0	38
15	390	39.0	26

[a] Based on amount of 2,4,5-T in 8 l water containing 200 mg of
273 micron Columbia LC carbon at 25°C. Equilibrium time less
than two weeks. (Extrapolation from Figure 6, Weber and Gould,
1966).

TABLE 9

Classes of Materials Related to the Effect of pH on Adsorotion (Hamaker and Thompson, 1972)

Class	Example	pK_A	Molecular Form		pH Effect
			Low pH	High pH	
Strong acid	Linear alkyl-sulfonates		Anion	Anion	Small
Weak acid	Picloram	3.7	Free acid	Anion	Large adsorption; pH approx. pK_A
Strong base	Diquat		Cation	Cation	Decrease at very low pH (18 N H_2SO_4)
Weak base	Ametryne		Cation	Free base	Increasing adsorption to pH approx. pK_A and then decrease
Polar molecules	Diuron		Un-ionized	Un-ionized	Small
Neutral molecules	DDT	Nil	Un-ionized	Un-ionized	Probably none

analytically determined organic carbon by a factor of 1.724) appears
to correlate best, for the most part, although roughly, with adsorption on soils of varying organic matter content than by other criteria. Adsorption coefficients calculated by the above authors on
a number of pesticides are useful for comparative purposes.

Pesticide Distribution Studies in Aquatic
Ecosystems in the Laboratory

The main purpose of bioconcentration studies is to determine
the distribution of residues, selective or otherwise, of a given
chemical over time in different organisms and segments of their
environmental surroundings which constitute typical or representative ecosystems.

The study of the partitioning and uptake of pesticides in
aquatic ecosystems can be as simple or complicated as one wishes
to make it, however, the simpler the system the more drastic each
change will make in the results of toxicity or residues studies.
The simplest aquatic system is where one organism is exposed in "pure"
water to one pesticide for a short period of time and where the addition of other variable parameters can be studied, one at a time.
Such variables might eventually include different temperatures;
types or concentrations of pesticides; ratios of organism weight or
size to water; degrees of aeration; light exposure; species; added
adsorptive surfaces such as soil, microorganisms, aquatic plants,
or animals; and added solutes such as inorganic salts, and organic
solutes.

The selective partitioning of a chemical from water to fish for
example represents a simple ecosystem in which little exists for the
pesticide to partition to besides the fish except for the glass
surface and air.

Effect of temperature change. The temperature affects residue
pickup if for no other reason than a change in rate of metabolism
of the organism. Reinert et al. (1974) found that the amount of
mercury (from methyl mercuric chloride) and DDT residues from DDT
accumulated from treated flowing water by yearling rainbow trout,
a cold water fish, increased as water temperature increased suc-
cessively from 5 and 10°C to 15°C (the latter temperature favored
by the trout) over a 12-week treatment period. Fish exposed to
methylmercuric chloride at concentrations of 234-263 parts per
trillion for 12 weeks at 5, 10, and 15°C accumulated 1.19, 1.71,
and 1.96 ppm; fish exposed to p,p'-DDT at concentrations of 133-176
parts per trillion accumulated 3.76, 5.93, and 6.82 ppm. Throughout
the period of exposure, the concentration factors (concentration of
contaminant in the fish/concentration in water) at each of the three
temperatures were far higher for p,p'-DDT than for methylmercuric
chloride.

Youngson and Meikle (1972) studied the effect of water temperature at 55° and 70°F on the rate of bioaccumulation of ^{14}C-DDT by
mosquito fish (*Gambusia*), a warm water fish species, in enclosed
aquaria. Fish weighing an average of 150 mg in 3 liters of water
reached a near equilibrium concentration of DDT within one day in
70°F water, while those in 55°F water had not reached equilibrium
in six days and had slightly more than one-half the residues of
those at 70°F (Table 10).

Effect of water flow. Lincer et al. (1970) compared the toxicity of endrin and DDT under static and flowing water bioassay
conditions. LC_{50}s for fathead minnows (*Pimephales promelas*), using
48 and 96 hour exposure periods, indicate a slightly higher value
during static tests compared to flowing water tests. From an applied nominal 40 ppm concentration, water concentrations of DDT
decreased from 33 to 7 ppb (partially due to volatility?) in static
water conditions, while applied concentrations of 40 ppb in flowing
water held between 22-34 ppb, in a 48 hour period of exposure. Thus
the increased toxicity of DDT under static water tests over those
in flowing water was not due to concentration effects. It is
reasonable to assume that decreasing oxygen concentrations and increasing fish metabolites such as ammonia, carbon dioxide, and
others may have enhanced the apparent toxicity of DDT during static
water exposure conditions.

Speed of pesticide uptake and clearance. The gills of fish
which strain large quantities of water, are extremely efficient
for rapid removal of DDT from water. Premdas and Anderson (1963)
found that after a five minute exposure to 1 ppm DDT (greater than
the water solubility of DDT), Atlantic salmon (*Salmo solar*) concentrated 1.53 ppm DDT in the liver and spleen and after one hour of
exposure concentrated 31 ppm.

Meikle et al. (1972) determined the speed of uptake of about
1 ppb of DDT in 16 liters of aerated water by mosquito fish at
77°F. Equilibrium was reached in about three to five days. The
DDT plus DDE concentration in pithed fish was about 13 to 14 thou-
sand ppb. At the end of the 17 day uptake study fish had reached
16,065 ppb. Residues in the latter group of fish, when placed in
fresh running water (300 ml/min change), dropped to 10,123 ppb,
6,381 ppb, and 4,167 ppb DDT + DDE, one, two, and seven days after
removal from the DDT treated water. From these studies it appears
that a quick 50% loss time clearance of DDT + DDE occurred in the
first two days followed by a much slower 50% loss time afterwards.

Cox (1971) studied the uptake, assimilation, and loss of [14]C
labeled DDT residues by a shrimp, *Euphausia pacifica*, in flowing
seawater at 10-12°C. Direct uptake of DDT from seawater at ppb
levels was rapid, reaching an equilibrium at 72 hours for small

TABLE 10

Effect of Temperature on DDT Uptake by Mosquito Fish
(Youngson and Meikle, 1972)

	Temperature, °F					
	55°			70°		
Exposure Time, Days	Conc DDT in Fish ppb	Conc DDT in Water ppb	CF[a]	Conc DDT in Fish ppb	Conc DDT in Water ppb	CF[a]
1	710	1.064	663	4290	0.917	4678
4	1698	1.022	1161	3887	0.930	4180
6	2690	0.986	2726	4709	0.913	5158

[a] $CF = \dfrac{\text{concentration in fish}}{\text{concentration in water}}$ = concentration factor

individuals (less than 3 mg day weight). Larger individuals took longer to reach equilibrium and longer to lose DDT when put in untreated water. Euphausid shrimp appeared to take up more DDT from water than from food.

Johnson et al. (1971) found ^{14}C labeled aldrin and DDT to rapidly accumulate residues of similar magnitude in the same species of aquatic invertebrates within 24 to 72 hours, using a continuous flow of water containing 0.1 ppb of the pesticides and no food. Bioaccumulation had not reached equilibrium in 72 hours indicating absorption in addition to adsorption.

Crosby and Tucker (1971) showed the principal uptake of DDT in *Daphnia magna* is through the carapace. DDT accumulates rapidly in 16,000 to 23,000-fold quantities from ppb suspensions of DDT in water within 72 hours. Absorption appeared to take place as well as adsorption since only one-half of the DDT was in moulted carapaces.

Saunders and Chandler (1972) using a continuous-flow system found four aquatic species of crustacea and four species of immature aquatic insects rapidly took up ^{36}Cl labeled polychlorobiphenyl (PCB-Aroclor[R] 1254) from ppb levels in water in four days to levels 1,700 to 47,000 times those in water.

   Chain of life organisms. Reinert (1972) studied the accumulation from water of low ppb concentrations of dieldrin in an alga (*Scenedesmus obliquus*), a daphnid (*Daphnia magna*), dieldrin in water. The amount of residue and time of accumulation were also proportional to the size (weight, surface area), life span and complexity of the distribution system of organism. For the smaller organisms, these factors favor greater ppm adsorption, quicker distribution time within organisms, short life span for accumulation, and less number of specialized organs and tissues for storage and accumulation of pesticides.

   Route of pesticide intake. Organisms in water are exposed to concentrations of pesticides from two main sources, those from the water directly through the skin, gills, cuticle, etc., and from oral dietary and water intake. In order to determine which was the most important route of residue uptake, Reinert (1972) measured the uptake of dieldrin by daphnids and guppies from water and compared these residue values to the uptake by daphnids and guppies feeding on food containing residues equal to those obtained by exposure to the same water concentration of dieldrin as those used in the water uptake studies. From these studies it appears that more dieldrin is picked up from water than from food by daphnids and guppies. Other studies have been made to determine whether the contact or oral intake of pesticides are most important. Examples of determining factors are the persistence, distribution, and concentration of the pesticide in water as contrasted to those in its food; the stability of the pesticide in the digestive system compared to water; the distribution coefficients of the pesticide between gill surface and blood versus the digestive system and blood of the fish; and as related to metabolism by the different routy of pesticide intake.

   Distribution of pesticides in fish tissues. Grzenda et al. (1970) showed the uptake, metabolism, and elimination of DDT, DDE, and DDE (jointly called DDT-R) in goldfish (*Carassius auratus*) fed C^{14} ring-labeled DDT in their diet. The dosage was equivalent to approximately 18 ppm of DDT in the diet of fish, averaging 21 g, at the rate of 3-4% of their body weight per week which is equivalent to 0.08 to 0.13 mg DDT/kg of body weight/day. These tests were conducted in flowing water at 15°C (57°F) and fish fed every other day for 192 days for a total of 842 nmoles DDT/gram wet tissue. Muscle tissue contained the least DDT-R (1.4 ppm) and mesenteric adipose tissue contained the most (42.6 ppm). Average lipid determinations for the tissues analyzed in the fish were 15% and ranged from 5 to 29% in muscle, and skin (subcutaneous fat removed) respectively. This data shows lipid to be widely distributed throughout all tissues. The mean relative distribution of DDT-R among various tissues of goldfish as compared to muscle tissue are shown in Table 11.

TABLE 11

Distribution of DDT-R Fed in Diets of Goldfish and Clearance Half-Life (Grzenda et al., 1970)

Tissue	DDT-R Tissue Residue Ratio to Muscle (After 192 Days Exposure)	Clearance Half-Life DDT-R (Days)
Muscle	1	30
Immature ovary	1.5	32
Blood	1.5	33
Gill	1.5	30
Skin	2.0	33
Gall bladder	3.8	29
Spleen	3.5	28
Stomach	3.5	30
Intestine	3.5	27
Kidney	4.0	31
Liver	5.0	25
Brain	5.0	30
Feces	6.0	29
Nerve	12.5	22
Testes	13.5	
Mesenteric adipose	23.0	34

The amount of DDT-R residues in various tissues of the fish
increased most rapidly in the first few weeks and then tapered off
but did increase throughout the study. Goldfish weights were
taken and probably increased throughout the study thus leaving
open the question of increased fat as they got older (and bigger).
However, the ratios of DDT-R in tissues as shown in Table 11 did
not change much throughout the 192 days of the study. The authors
stated that "no correlation was found between DDT concentration
and the lipid content of various tissues," although certainly the
highest concentration is in mesenteric fat. Half-life in days were
calculated from a plotted regression slope for the various tissues,
and the half-lives were found not to vary greatly (22-34 days) as
shown in Table 11. This data indicates that the fish metabolizes
DDT-R from most parts of its body rather evenly which is in keeping
with the rather even distribution of lipids in its body and that
the half-life of DDT-R whether in fat or other tissues may be dependent upon the speed of metabolism such as availability to the
blood stream, difference in types of lipids and locations of fat
and other types of DDT-R bearing tissues.

Rodgers and Stalling (1972) studied the dynamics of 0.3 or 1
ppm of the ^{14}C-labeled butoxyethanol ester (BE) of 2,4-D in the
organs of three species of fed and fasted fish. Maximum ^{14}C residue concentrations were observed in most organs of fed fish within
one to two hours of exposure and within one to eight hours of exposure in fasted fish. Concentrations of (BE) ester of 2,4-D
hydrolyzed in static water from 1 ppm to 0.1 ppm in 90 hours while
in the presence of fish 99% deesterification took place in 24 hours.
Only 2.5 ppm 2,4-D was lost from water probably due to volatilization while in the form of the ester and to sorption on the fish.
Highest concentrations of ^{14}C residues occurred in the bile and
kidney, liver, and gill tissues while blood, brain, muscle, stomach
had low amounts. Since ^{14}C residues are indicators of the ester,
it appears that the fish rapidly deesterified 2,4-D (BE) and in
spite of continued exposure to essentially the same concentrations
of ^{14}C in the water throughout the exposure bile was the only significant source of ^{14}C after 72 hours. This study shows that
2,4-D or 2,4-D (BE) did not accumulate in the various tissues of
fish but was rapidly excreted via the bile.

Correlation of octanol:water partition coefficients with bioconcentration factors in fish. Branson et al. (1974) have studied
the rate constants of the uptake and clearance of 2,2',4,4'-tetra-
chlorobiphenyl to aid in the prediction of these phenomena with
other stable organic compounds. Residue samples were taken at
various time intervals from muscle tissues of rainbow trout (Salmo
gairdneri) held at 53°F in flowing water treated at a constant
rate with various compounds. It was shown that short-term (five
days) uptake of the pesticides was a reliable indicator in predicting
longer term (42 days) uptake. Steady state uptake data from other

compounds having a wide range of physical and chemical properties
were calculated from short term uptake and were compared with concentrations in water as shown in Tables 12 and 13 to obtain the
bioconcentration factor (BF).

Neely et al. (1974) obtained partitioning coefficients (PC)
of these compounds studied by Branson et al. (1974) in n-octanol
over water and developed a regression equation from this data for
the prediction of bioconcentration factors. The developed equation
is the log of the bioconcentration factor = 0.56 log of the partition coefficient + 0.124. This equation was used to predict the
bioconcentration factor of three other chemicals with a reasonable
degree of fit with experimental data (Table 12). The assumption
of correlation between the n-octanol:water partition coefficient
and the bioconcentration factor between fish and water seems reasonable to within one to two orders of magnitude. Part of the
difference between the two values is narrowed if the fairly low
fat content of muscle (2-4%) is taken into account. In addition,
the bioconcentration of fat soluble compounds in fat depots in fish
will result in higher concentrations there than in the whole fish
(which may rarely contain as much as 10% fat in its body weight).

Correlation of pesticide concentration and biomass; adsorption
and absorption. Kenaga (1972) discussed the relationship of the
magnitude of pesticide residue accumulation to the changing biomass
of organisms or changing concentrations of pesticides. It was
pointed out that bioconcentration factors could vary greatly by
changing the ratio of biomass to water volume containing the pesticide and the ratio of surface area to weight ratio of organisms in
water.

A precise study of the relationship was carried out by Johnson
and Kennedy (1973) who studied the accumulation of residues of C^{14}-
labeled DDT and methoxychlor by two species of bacteria by varying
the ratios of biomass to pesticide concentration in water. Total
cellular residues of both pesticides in water rose linearly with increasing concentrations of 0.5 to 5.0 ppb (10×1^{-9}) while keeping
the bacterial mass and water volume constant during a two-hour
exposure period. Also biomagnification factors were reduced nearly
proportionally to the increase in microbial biomass when DDT or
methoxychlor concentrations were kept constant during a four-hour
exposure. The data show that biomagnification ratios of pesticide
residues on bacteria to those in equal volumes of water are variable, depending on a change in either biomass or concentration of
pesticide. It is therefore important to make use of a standardized
balanced ecosystem as a test tool.

It is also important in evaluating the environmental toxicity
of a compound to determine whether residues soon reach a maximum
concentration or whether residues increase over period of days,

TABLE 12

The Relationship of Partition Coefficients and Bioconcentration Factors in Rainbow Trout
(Neely et al., 1974)

Compound	Partition Coefficient (PC)	Bioconcen-tration Factor (BF) [c]	Ratio PC:BF	Residues in Fish (ppm) Uptake From 1 ppb in Water
Tetrachloroethylene	180[a]	39	5	0.039
Carbon tetrachloride	440[b]	17	23	0.017
p-Dichlorobenzene	2,400[b]	230	10	0.23
Diphenyl ether	15,900[b]	200	79	0.20
2-Biphenylyl phenyl ether	460,000[a]	550	832	0.55
Hexachlorobenzene	1,700,000[a]	8,500	195	8.5
2,2',4,4-Tetrachloro-diphenyl	5,200,000[a]	12,300	426	12.3

[a] Calculated values, Hansch et al., 1972. (Accurate to only two significant figures.)
[b] Leo et al., 1971.
[c] The log of BF can be estimated from the equation $0.56 \times \log$ of PC $+ 0.124$.

PC = partitioning coefficient between n-octanol and water.
BF = bioconcentration factor between fish and water.

weeks, or months. In many studies, the residues are often not measured (for background levels from unknown sources) before the application of the chemical, nor immediately afterwards. In addition
to this, two sequential phases of residue bioconcentration are not
differentiated: namely 1) physical adsorption to surfaces which
occurs almost immediately, (related to surface area/volume measurements of the organism); 2) long term organism penetration (absorption). This differentiation is important since continued build-up
of residues in organisms over time is mostly due to absorption and
redistribution within the organism and is a function of its metabolism. Adsorption factors may result in residues which constitute
a base-line of residues to be subtracted from total (absorbed and
adsorbed) residues, thus helping to determine those residues due to
penetration of the chemical into the organims over a period of
time (i.e., bioconcentration because of absorption).

   ^{14}C-labeled DDT and methoxychlor by two species of bacteria was
rapid; 80 to 90% of the 24 hour residues were reached in 0.5 hour.
Radiometric methods were used to determine the total residues (both
adsorption and absorption) by bacterial cells. After the insecticide-
exposed microbial cells were washed with pesticide-free water, DDT
residues decreased 55% in *Aerobacter aerogenes* and 70% in *Bacillus
subtilis*, whereas the methoxychlor level decreased nearly 75% in
both organisms. Subsequent washing did not further reduce the insecticide residue, indicating a more permanent binding for the
remaining residue whether adsorbed or absorbed.

   A more complicated aquatic ecosystem containing soil and plant
and animal organisms was developed for the purpose of differentiating
between adsorbed and absorbed pesticides by Smith et al. (1972) who
included measured amounts of each ingredient and kept a constant
ratio of sorptive materials. The system consisted of an all glass
aquarium containing the following:

   1. 2 inches of sandy loam (1.6% organic matter) soil in the
bottom of the tank;

   2. 40 liters of city water (dechlorinated);

   3. 40 grams of floating plants (duckweed, *Lemna minor*);

   4. 40 grams of submerged plants (fanwort, *Cabomba caroliniana*);

   5. 200 daphnids (*Daphnia magna*);

   6. 200 grams of mixed algae culture;

   7. 40-48 grams of fish (goldfish, *Carassius auratus*), approximately 2 grams each.

The system was aerated at the rate of 200 ml of air/minute through standard aquarium stones and the water temperature was maintained at 65°F. The system was illuminated with incandescent lamps (G.E. Par 150 Flood Lamps) mounted 18 inches above the tanks.

The distribution of high specific activity ratio-active C^{14}-labeled dalapon sodium salt in this aquarium system containing 5 ppb dalapon in water was studied over a three-day exposure period. A second identically furnished aquarium was used except that instead of using live organisms, the plants were killed by autoclaving and the animals were killed by freezing them in dry ice prior to putting them in the aquarium. The difference with time between residues from organisms in the two aquariums were essentially that due to adsorption and absorption in the first aquarium minus that due to adsorption in the second aquarium. Concentrations of dalapon in water remained at 5-6 ppb throughout the exposures, showing little or no volatility from water or overall adsorption from water by organisms. Fish appeared to contain low amounts (ppb) on or in tissues of both dead and live whole fish. Snails and *Daphnia* sp. contained residues less than that of the surrounding water environment. Duckweed, *Lemna minor*, contained about 40-50 ppb when dead and up to about 180 ppb, while alive (before dalapon killed the roots). *Cabomba caroliniana* contained about 20-50 ppb when dead, but around 780 ppb when alive and appeared still be to absorbing residues after 72 hours exposure (Table 13). In the case of dalapon treated dead plants showed adsorptive bioaccumulation while dalapon treated live plants showed both adsorptive and absorptive bioaccumulation. Animal species treated with dalapon showed neither type of bioaccumulation from water.

Smith et al. (1972) passed 80 ml of air per minute per liter of water in a 10 liter aquaria containing pesticides to study codistillation. Under such aeration conditions no dalapon at 32 ppb disappeared from water over a 120 hour period while in an equivalent system 65 ppb of DDT decreased to <1 ppb in water in the same period of time. DDT codistilled with water while dalapon did not.

Smith et al. (1972) also studied the ^{14}C distribution of 5 ppb of ^{14}C radio-labeled chlopyrifos in the same type of aquarium ecosystem used for dalapon. The comparative data are shown in Table 14, and indicate that a great proportion of the residue accumulation on plants and fish may be caused by physical processes such as adsorption. Both dead and live fish and plants acquire residues of chlorpyrifos several hundred times the concentration found in the water within eight to thirteen hours which then decline rather rapidly apparently due to metabolism of chlorpyrifos.

Dead plants acquired higher residues than live plants, while live fish acquired higher residues than dead fish. Water residues of chlorpyrifos declined to <0.1 ppb even though chlorpyrifos is

TABLE 13

Uptake of Dalapon Sodium Salt by Fish and Plants, Dead and Alive From Water
Treated With 5 ppb ^{14}C Dalapon (Smith et al., 1972)

Organisms or Media--Residues of Dalapon

Hours After Application (Approximate)	Goldfish		Duckweed		Fanwort		Water ppb
	Alive ppb	Dead ppb	Alive ppm	Dead ppm	Alive ppm	Dead ppm	
1	3	6	0.05	<0.05	<0.05	<0.05	6
5	1	18	0.05	<0.05	0.1	<0.05	6
12	4	6	0.05	0.05	0.2	0.05	6
24	3	11	0.1	0.05	0.2	0.05	5
48	8	13	0.1	0.05	0.4	0.05	5
72	-	--	0.2	0.05	0.8	<0.05	6

TABLE 14

Uptake of Chlorpyrifos by Fish and Plants, Dead and Alive, From Water
Treated with 5 ppb Chlorpyrifos (Smith et al., 1972)

Organism or Media--Residues of ^{14}C Chlorpyrifos

Hours After Application (Approximate)	Goldfish		Duckweed		Fanwort		Water
	Alive ppm	Dead ppm	Alive ppm	Dead ppm	Alive ppm	Dead ppm	ppm
8	1.6[a]	--	--	2.3[a]	<0.65[a]	1.7[a]	2-3
10	--	0.35[a]	--	--	--	--	--
13	--	--	0.9[a]	--	--	--	2
72	0.75	0.15	<0.1	0.2	0.55	1.0	--

[a] Peak concentration measured.

soluble at 400 ppb showing that the partitioning coefficient of
chlorpyrifos between organic matter and water is highly in favor
of organic material, whether it be of plant or animal origin.

A similar aquatic ecosystem with soil, plants, and animal organisms in water containing 50 ppb of ^{14}C chlorpyrifos showed a drop
in concentration of chlorpyrifos in water to 6 ppb in ten hours,
which then leveled out to 4 ppb after 120 hours (Smith et al., 1966).
Ten hours after application, the ^{14}C metabolites of chlorpyrifos,
apparently released from fish, plants, etc. began to increase rapidly
in water and rose to 19 ppb, giving a total ^{14}C radioactivity equivalent to 23 ppm of chlorpyrifos metabolites after 120 hours. Metcalf
et al. (1971) developed an aquarium model ecosystem involving the
treatment with pesticides of plant feeding insects and plants growing on a sloped sandy soil which tapered into and below water. The
larvae, snails, and fish, the components of a change-of-life ecosystem
much as in a natural ecosystem. From the use of this test method
Kapoor et al. (1970) were able to show that DDT and methoxychlor
indeed go from the land phase to water and organisms in it and concentrate from very small quantities in water to the highest concentration in snails and fish. DDT was shown to metabolize to DDD and
DDE, both of which also bioconcentrated in quantities of three to
four orders of magnification from water concentrations, similar to
DDT.

Isensee et al (1973) used a modified Metcalf model ecosystem
to study the distribution of ^{14}C labeled aklyl arsenicals. Cacody-
lic acid (CA) and dimethylarsine (DMA) were applied directly to
water containing known numbers or weight of microorganisms, algae,
Daphnia magna, mosquito fish, and algae, *Oedogonium carciasum*.
Exposures of these organisms were for 3, 29, 32, and 32 days, respectively. Lower food chain organisms (algae and daphnids) bioac-
cumulated more CA and DMA than did higher food chain organisms
(snails and fish). A gradual increase in the biomass, primarily
algae, over 32 days largely accounted for a gradual loss of CA and
DMA from solution, algae being 67 to 74% of the biomass in this
ecosystem. It was concluded that large differences in bioaccumulation ratios of CA and DMA for the organisms may be due in largest
part to adsorption and to larger organisms having the smaller surface
area to mass ratios. Bioaccumulation ratios were based on [14]C
counts per minute of dry organism tissue compared to water. The
results of these tests with CA and DMA are shown in Table 15. The
highest bioaccumulation ratios of CA were shown to be related to
the lowest water solution concentrations (range--0.1, 1.0, 10 ppm).

These results obtained by Isensee et al. (1973) on bioaccumulation ratios using daphnids were compared with those obtained by
Johnson et al. (1971) for correlation with various chemical structures (Table 23). There is a paucity of comparative data on bioaccumulation, testing a number of widely different chemical structures
on the same organisms, using the same test methods.

TABLE 15

Bioaccumulation Ratios of Residues of Various ^{14}C Labeled
Compounds Between Dry Tissues of *Daphnia magna* and Water

	Concentration in Water		
Compound	Beginning of Test, ppb (Days)	End of Test, ppb (Days)	Bioaccumulation Ratio at End of Test
DDT	0.080(0)	?(3)	114,000[a]
Aldrin	0.060(0)	?(3)	141,000[a]
Cacodylic acid	10.6(1)	6.1(32)	1,658[b]
Dimethyl arsine	7.0(1)	3.9(32)	2,175[b]
Endothall			150[b]
2,4,5-T	210(2)	126(28)	217

[a] Residues in dry weight of tissue compared to concentrations in water at beginning of test. Tested in continuous flow system at 21°C. Johnson et al., 1971.

[b] Residues in dry weight of tissues compared to concentrations in water at end of test. Tested in a static water system at 22°C. Isensee, 1973.

Distribution Studies From Field and
Microplot Applications of Pesticides

Kenaga (1972) concluded that the principal metabolites of DDT in the environment are DDD and DDE which appear to be as persistent or more so than DDT. They also appear in all segments and components of the environment where DDT occurs. Birds and fish appear to have particularly high residues of DDE. DDT, DDD, and DDE partition in favor of fat tissues and disfavor water. Plants do not appear to pick up residues when grown in soil containing DDT, DDD, or DDE. Organic slicks which occur in ocean water may concentrate DDT and especially DDE, as well as particulate carbon and dissolved organic carbon. Certain organic solvents present in ocean water may act as cosolvents for DDT in water or the organic slicks. Variations in the amounts of DDT in various anatomical fat parts of an individual bird may be due to differential rates of deposition, and metabolism of the fat containing DDT.

Reinert (1970) pointed out that as fish grow, their percentage of fat in the body increases, and that in Lake Michigan lake trout, *Salvelinus namaycushi*, the percentage of fish oil is directly proportional to whole fish DDT residue values. Reinert and Bergman (1974) show that DDT-R residues in Lake Michigan coho salmon, *Oncorhynchus kisutch*, have a marked decrease in oil content from 13.2% in August to 2.8% in January, losing their adipose tissue in October, and that DDT-R concentrations were redistributed in high quantities in the loin and brain. Also, DDT-R residues were redistributed in tissues of spawning run fish.

It is of interest to contrast the distribution of DDT to the distribution of compounds of much greater water solubility such as most of the herbicides.

Smith and Isom (1967) reported on the effects of a massive herbicidal test in TVA reservoirs in Tennessee and Alabama. A granular formulation containing 20% 2,4-D equivalent, in the form of the butoxyethanol ester, was applied to waters averaging about 20 feet in depth for the control of European watermilfoil (*Myriophyllum spicatum* L.).

The low concentration of 2,4-D found in water in comparison with the concentration expected if it had all gone into the water phase was probably due to slow release of the 2,4-D from the granular formulation and to sorption of the 2,4-D ester on the bottom of the reservoir (Table 16). In contrast to DDT, 2,4-D does not accumulate in fish and only slightly in mussels.

While DDT does not accumulate in plants, 2,4-D did accumulate sufficiently (8 ppm) for the target herbicidal effect, within 24 hours. This concentration is at least 25-fold over than in water.

Cope et al. (1970) studied the distribution of 2,4-D as a herbicidal formulation (Esteron[R] 99) in the form of PGBE esters (propylene glycol butyl ether esters) in representative components of man-made outdoor ponds. Residues of 2,4-D were determined in water, bluegills (*Lepomis machrochirus*), pond weeds (*Potamogeton nodosus*), and bottom sediments from the ponds over a period of time. Ten ppm applied in ponds of the depth described is equivalent to approximately 82 lb/A of 2,4-D PGBE or 50 lb/A of 2,4-D acid equivalent, a high dosage. Highest residues of 2,4-D were found in plants, followed by water, mud, and fish (Table 17). No attempt was made to determine whether 2,4-D residues were present in the acid or ester form although esters of 2,4-D are known to be hydrolyzed readily in

TABLE 16

Comparative Residue Values in Various Aquatic Environmental
Segments in the Watts Bar Reservoir Due to 2,4-D Applied in
a Granular Formulation at the Rate of 100 lb (AE)/A of Water
(Adapted from Smith and Isom, 1967)

Component Analyzed	Time After Application	ppm Residue, 2,4-D BBE[a]
Water	Immediately	2.67[b]
Water (not filtered)	1 hour	0.037[c]
Bluegull (*Lepomis machrochirus*)	72 hours	<0.14
	50 days	<0.14-0.15[c]
Mussel (*Ellipto crassidens*)	96 hours	0.70[c]
Water milfoil (*Myriophyllum spicatum*)	24 hours	8.26[c]
Mud	96 hours	0.38-56.0
	10 months	0.24-58.8[c]

[a] The analytical technique does not distinguish between the acid
(AE) and the butoxy ethanol ester (BEE) forms of 2,4-D. Undoubtedly the acid is present, but in unknown amounts. Detection
limit in animals and plants is 0.14 ppm.

[b] Theoretical average possible in water 20 ft deep if 2,4-D (BEE)
is evenly mixed and immediately released from a 20% 2,4-D acid
equivalent (AE) granular formulation.

[c] Maximum value measured at any time after application.

animal organisms. The original distribution of the 2,4-D was in the
form of the ester which is highly lipid soluble and would have a
different partitioning coefficient than the acid. See Table 18
for similar changes with 2,4,5-T.

Walker (1971) reported that extracts of tissues from bluegill
and channel catfish exposed for two hours to 2,4-D butoxyethanol
ester (BEE) showed it to be present only in the liver. All organs

TABLE 17

Comparative Residue Values in Various Environmental Segments in Tishomingo, Oklahoma, Ponds
Due to 2,4-D Applied in an Emulsifiable Formulation of 2,4-D as the PGBE Ester
(Adapted from Cope et al., 1970)

Pond Component Analyzed	Days After Application	Residues of 2,4-D (Acid Equivalent) (ppm)				
		Dosage Applied - 2,4-D (ppm)				
		10	5	1.0	0.5	0.1
Water[a] (not filtered)	13	8.0	4.0	0.2	1.0	0.06
Pond weed	2	50	3.8	--	--	--
	8	6.0	3.5	0.9	0.5	ND
	15	5.0	2.3	1.0	0.3	--
	94	0.1	0.1	--	--	--
Bottom sediments (depth?)	2	3.0	1.1	--	--	--
	9	1.7	0.3	0.2	0.1	ND
	17	1.4	0.5	0.1	0.1	--
	44	0.1	--	--	--	--
Bluegill (whole body)	1	2.0	1.0	ND	ND	ND
	3	1.6	0.3	ND	ND	ND
	14	ND	ND	--	--	--
Lb 2,4-D (as PGBE) Applied/A (Theoretical concentration in water averaging 3 feet deep)		81.6	40.8	8.16	40.8	0.82

[a] Bioassayed with cucumber seed. Other residue values determined by chemical analyses.

ND = not detectable

TABLE 18

Comparative Residue Values of Diuron in Various Aquatic
Components at Tishomingo, Oklahoma, When Treated With
A 80% Wettable Powder Formulation (McCraren et al., 1969)

Component Analyzed	Days After Application	Residues of Diuron (ppm)		
		Dosage Applied (ppm)		
		3.0	1.5	0.5
Water	1	0.01	ND	0.01
	4	0.24	0.20	ND
	21	0.04	ND	0.01
Muds	3	4.7	1.2	0.25
	10	2.6	3.5	0.32
	24	1.5	0.4	0.09
	80	0.19	0.4	0.27
Aquatic rooted vegetation	4	20.4	2.4	4.5
	11	1.9	3.4	3.8
	18	3.9	ND	ND
Fish--Bluegills	3	40	19	20
	14	37	6	0.8
	28	8	8	1.0
Lb diuron applied/A (theoretical in water averaging 3 ft deep)		24.5	12.2	4.1

contained 2,4-D, indicating that the ester linkage of 2,4-D BEE is
rapidly broken by hydrolysis to 2,4-D. 2,4-D BEE hydrolysis proceeded much more slowly in water without fish. Maximum body and
tissue residues of 2,4-D are related to the length of time of availability of the ester, which is more lipoid soluble (penetrates the
gills better) and less water soluble than 2,4-D. Residues decline
as the ester is hydrolyzed. Whole body residues were 7 to 55 times
greater than water one to six hours after exposure. Hydrolysis of
the ester to the acid apparently takes place after passing through
the gill membrane, since detectable quantities of 2,4-D acid were
not concentrated in fish from water through fish gills. Rats fed
2,4-D acid rapidly transported the acid to various organs and was
mainly excreted in the urine, which is similar to the excretion

rate of 2,4-D in fish. Mammals and birds hydrolyze esters of 2,4-D
in their gut. Residues of the 2,4-D, dimethyl amine salt, were
measured in various tissues of three species of fish and were not
found to significantly exceed environmental water concentrations of
2,4-D. Two unidentified metabolites were present, one polar and
one relatively non-polar. In all three species the highest concentration of residues were in the bile which would point to excretion
of water soluble material by the gall bladder (Walker, 1971).

McCraren et al. (1969) determined the distribution of residues
of diuron in pond environments treated with an 80% wettable powder
formulation applied at 0.5, 1.5, and 3 ppm (by weight) of diuron
in water. Pounds per acre data were calculated on the basis of
water 3 feet deep. Results in Table 19 show that the highest residues occurred in fish and aquatic vegetation. Water residues were
not detectable after 21 days. While residue values were not given,
invertebrate populations increased similar to the controls and good
control of filamentous algae and vascular aquatic vegetation was
demonstrated during the observation period. Changes in the biological oxygen demand (BOD) which occurred in all treatments for at
least 30 days post-treatment was attributed to the herbicides rapid
destruction of vascular aquatic plants and resultant decay.

Cope et al. (1969) studied the distribution of dichlobenil in
various segments of two pond ecosystems using two different formulations (Tables 19 and 20).

A wettable powder formulation of 40, 20, and 10 ppm (= 326,
163, and 82 lb/A, respectively) of dichlobenil gave quick release
of concentrations of dichlobenil to water resulting in peak concentrations at six hours (or less). Quick partitioning of dichlobenil to fish occurred as shown by residues, and probably to plants
(since they were soon killed).

A granular formulation containing 4% dichlobenil, applied at
10 lb/A (= 0.58 ppm) of dichlobenil gave slow release to water
which peaked in residue concentration 34 days after application.
Residue concentrations were shown to be highest in algae (probably
because of high surface to mass ratio from adsorption). Dichlobenil
residues remained in the environment for many days in both tests.
Differences in the formulation of dichlobenil obviously controlled
and accounted for the large variation in the rate of release of
dichlobenil to the surrounding water and environment.

It may be concluded that the distribution of 2,4-D and derivatives, dichlobenil and diuron, which are herbicides, differ from
that of DDT-R, dieldrin, etc., especially by the fact that some of
the highest residues are held by aquatic plants. Formulation plays
a part in the release of pesticides, some of which are designed for
slow release to water.

TABLE 19

Comparative Residue Values of Dichlobenil in Various Aquatic
Components of Ponds at Tishomingo, Oklahoma, When Treated
With a 50% Wettable Powder Formulation (Cope et al., 1969)

Component Analyzed	Time After Application	Residues of Dichlobenil (ppm) Dosage Applied (ppm)		
		40	20	10
Water[a] (not filtered)	6 hours	130	50	80
	24 hours	2.0	1.6	0.35
	47 days	0.15	0.02	ND
Mud[b] (1-1.5 inches deep)	39 days	5.65	0.18	0.91
	53 days	22.95	2.31	0.12
Fish (whole body)	3 days	24.5	129	98
	11 days	--	4.6	--
	52 days	1.65	0.19	0.32
Lb dichlobenil applied/A (theoretical in water averaging 3 ft deep)		326.4	163.2	81.6

[a] Bioassayed with cucumber seed.

[b] Values highly variable.

PHYSICAL AND CHEMICAL PROPERTIES RELATED TO DISTRIBUTION OF PESTICIDES IN THE ENVIRONMENT

As discussed in the introduction, all chemicals in the environment are involved in distribution within or between organisms and
their external environments involving gas, liquid, and solid phases.
These phases are distinct and a chemical thus is forced through a
series of "choices" in its immediate environment resulting in distribution ratios of the chemical in one phase(s) compared to that
in another phase(s), otherwise known as partition coefficients.
Some of the simple partition coefficients occurring outside organisms
are commonly air:water; water:soil; water:animal; water:plant;
soil:plant; soil:air; and within organisms, water:fat; water:protein;

TABLE 20

Comparative Residue Values of Dichlobenil in Various Aquatic
Components of Bax Pond hen Treated With a 4% Granular
Formulation at the Rate of 250 lb/A (10 lbs Dichlobenil/A)
(Cope et al., 1969)

Component Analyzed	Time After Application	Residues of Dichlorbenil (ppm)
Water (theoretical)[a]	immediately	0.58[a] (applied)
Water (actual) (not filtered)	4 hours	0.05
	24 hours	0.06
	34 days	0.24
	92 days	0.05
	189 days	0.001
Mud (1-1.5 inches deep)	6 hours	9.2
	2 days	12.9
	4 days	6.5
	16 days	3.7
	62 days	2.0
	166 days	0.002
Chara (alga)	6 hours	45
	2 days	78
	8 days	22
	16 days	5
Filamentous algae (average)	6 hours	312
	24 hours	508
	2 days	154
	8 days	18
	34 days	5
	151 days	0.1
Fish (range of 4 species, whole body)	1 day	1.2-2.1
	4 days	2.8-3.9
	34 days	4.5-8.0
	92 days	0.5-0.8

[a] Based on average water depth of 5.5 feet.

water:sugar; water:air; etc. Of course, these are rarely simple
binary systems since all of the other ingredients in each phase
also interacts according to its properties.

In the discussion on natural bodies of water, blood, and plant
sap, it becomes apparent that all contain chemicals with similar
functional groups but in varying proportions and with many distinct
differences.

Goring (1972) has shown the comparative sorption of lipophilic,
hydrophobic molecules such as DDT from water by hydrocarbon containing organisms such as aquatic plants and animals and by soil organic
matter to be similar to within one or two orders of magnitude (0.32-
14.1 x 10) (Table 21). The variation in sorption is partly due to
the way the values were calculated (using total organic matter in
soil and whole body residues for animals to correlate distribution
coefficients).

The principal difference in the way a compound partitions is
apparently related to whether it is polar (as judged by water solubility) or non-polar (as judged by fat solubility). Some representative levels of the water solubility of simple organic compounds
found in nature and industrial wastes are shown in Table 22 (Leo,
1972) and show a range of water solubility in the order of three to
four magnitudes. Thus, it is not surprising that some pesticides
partition in certain tissues to at least that order of magnitude.
The choice of n-octanol, a compound midway in the list, as a solvent
for determining representative partitioning between fat and water
is probably as good as any. Non-polar compounds like DDT appear to
partition in n-octanol and water similar in magnitude to the partition coefficient of DDT in the fat tissues of aquatic animal organisms over that in DDT contaminated water. Calculation of weight
ratios of pesticides in water to air or organic matter to water have
been proposed to estimated relative partition coefficients for pesticides (Goring, 1967) as a guideline to how the pesticide would act
in nature from the standpoint of leaching, volatility, and sorption
to various ecosystem components.

Such coefficients cannot be construed as representing the absolute distribution coefficients that actually occur in nature since
these are governed by an enormous range of factors. Moreover, these
relative distribution coefficients should be considered as rather
crude, but serviceable indications of pesticide behavior rather than
precise tools for prediction.

Necessarily, a number of assumptions in distribution calculations must be made for the purpose of uniformity. For example, in
air to water relationships, one must assume the ratio of weights to
be in equilibrium, in equal volumes of air and water at the same
temperature, and that the chemical follows the gas laws and that the

TABLE 21

Comparative Sorption of DDT From Water by Aquatic Organisms
and Soil Organic Matter (Goring, 1972)

Adsorbing Material	Exposure Period (Days)	Distribution[a] Coefficient
Soil organic matter	4	14.1×10^{4} [b]
Golden shiners (*Notemigonus crysoleucas*)	1	0.32×10^{4}
	6	3.59×10^{4}
	9	4.26×10^{4}
	12	5.53×10^{4}
	15	5.36×10^{4}
Green algae (*Chlorella pyrenoidosa*)	2	1.10×10^{4}
Daphnia magna	2	11.4×10^{4}
Mosquito fish (*Gambusia* sp.)	2	1.93×10^{4}

[a] [Concentration in adsorbing material (dry wt. basis)]/(Concentration in water) = distribution coefficient.

[b] Transformed from a soil carbon to a soil organic matter basis, $K_{OC} = K_{OM} \times 1.74$.

concentration in air at the listed vapor pressure is assumed to be
in equilibrium with the concentration in water at maximum solubility
in water. There are quite a few assumptions in nearly every part
of this equation which may be difficult to fulfill. Equilibrium
is rarely acheived in nature due to the constant flow of water and
air and impurities in each. What is the true solubility of a pesticide in natural water which includes many cosolvents (usually in
small amounts) and solids which absorb pesticides? How does the
vapor pressure of a compound help predict its volatility? Does the
pesticide form an azeotrope with water? Is the pesticide volatilizing from an oil slick floating on water, or from floating particulate matter to which it is strongly adsorbed?

TABLE 22

Comparison of Water Solubility of Various Organic Solvents
(Leo, 1972)

Lipoid Phase	Solubility in Water Moles H_2O ($\times 10^3$)
1. n-Butyl alcohol	9940
2. Cyclohexanol	6510
3. 2-Butanone	5460
4. 2-Pentanol	5320
5. Primary pentanols[a]	5000
6. Cyclohexanone	4490
7. Octanol	2300
8. Ethyl acetate	1620
9. Methyl isobutyl ketone	950
10. Oleyl alcohol	712
11. Ether	690
12. Isopentyl acetate	456
13. Nitrobenzene	180
14. Oils	72.5
15. $CHCl_3$	68.4
16. Benzene	26.0
17. Toluene	25.6
18. Xylene	18.8
19. CCl_4	10.0
20. Heptane	3.3
21. Cyclohexane	2.5

Increasing Lipophilicity (arrow pointing downward, left margin)

[a] Average of n- and iso-amyl alcohols.

The partitioning of pesticides on soil is predicated on the fact that organic matter is the principal absorbent in soil and so the amount of organic matter is the main basis of the degree of partitioning in each soil. In determining the amount of organic matter in soil analytically the total soil carbon is usually determined. Based on the average carbon, hydrogen, nitrogen, and oxygen ratios of compounds in soil organics, a factor of 1.74 is applied to total carbon to convert it back to a soil organic matter basis. The mineral portion of the soil although important in some cases, is neglected for these coefficients. From these assumptions the soil organic matter to water partition coefficients are calculated by obtaining the ratio of chemical sorbed per gram of organic matter in equilibrium with the amount of chemical dissolved per gram of water. Again there are many qualifying conditions and many conditions left out of the equation. Soil to air partitioning of many pesticides is minimal in the absence of water (dry soil), and so the volatility of the pesticide from the available surface water of soil particles, and its release from soil sorption to water.

The partitioning coefficient (PC) for pesticides discussed here so far deal mostly with adsorption. When partitioning into organisms the phenomena of adsorptive mechanisms are important. While water is the main transport fluid for plants and animals, it must be controlled by membranes and cellular barriers in order to direct its use to the benefit of the organism. Osmosis and diffusion of gases and liquids may be the principal routes across such barriers. Little work has been done to measure the adsorptive and absorptive characteristics of such barriers, however, it is obvious that certain pesticides readily cross these barriers and get into the blood stream or plant sap while others are very selectively absorbed. DDT, for example, does not readily cross the aqueous plant root barrier nor human skin barrier but does readily cross lipid gill membranes of fish, and insect cuticles. The reasons for these effects are complex in detail but simple in broad outline as related to solubility, chemical stability, and volatility of the pesticide and its partitioning to tissues of similar or diverging chemical, physical, and biochemical properties.

From the previous discussion, it is seen that properties such as vapor pressure, molecular weight, solubility in water and various distribution coefficients are valuable in predicting the partitioning of pesticides in various situations. These properties known for compounds discussed in this paper are shown in Table 23.

Vapor pressure varies greatly among these compounds from below the vapor pressure of water to 6×10^{-7} mm Hg. Molecular weight varies from 84.9 to 381 and is an indicator of volatility, solubility, and size of the molecule. Solubility in water varies from 1.2 ppb to over 800,000,000 ppb. Partition coefficients in octanol:water varies from 0.17 to 5,248,000. An interesting factor regarding the

TABLE 23

Comparison of Vapor Pressures, Molecular Weight, Solubility in Water, and Various Partitioning Coefficients of Pesticides and Related Chemicals

Compound	Vapor Pressure 25°C (mm Hg)	Mol. Wt.	Sol. in Water 25°C ppm	n-Octanol: Water Partition Coefficient	Isooctane: 80% Acetone Partition Coefficient[a]	Soil Organic: Water Partition Coefficient[b]
A) HERBICIDES AND RELATED MATERIALS						
dalapon	--	143	>80%[c]	5.7[c]	--	--
dalapon sodium salt	--	164.9	90%	<1.0[c]	--	--
2,4-D	6×10^{-7c}	221	890	--	0.19	22-32
2,4-D butyl ester	8.86×10^{-6d}	277	<890	--	--	
2,4-D BEE	4.5×10^{-6e}	321	<890	--	--	
2,4-D PGBE	3×10^{-6e}	325	<890	--	--	
2,4,5-T	$<6.46 \times 10^{-6c}$	255.5	280[d]	7.1[f] (pH 5) 0.53[f] (pH 7) 0.17[f] (pH 9)	0.22	37-42
2,4,5-T butyl ester propyl ester[d]	7.46×10^{-5}	312.5	<280	64,000	--	

Compound	Vapor pressure	MW				
picloram	6.16×10^{-7} (35°C)[c]	241.5	430	1.8[c]	--	6-24
dichlobenil	5.5×10^{-4} (20°C)[g]	172	20	--	1.5	164
diuron	0.31×10^{-5} (50°C)[g]	233.1	42[g]	--		94-485
endothal	--	230	21%[g]	--	0.01	
dimethyl arsine	--	106	soluble	--	--	
cacodylic acid	--	138	soluble[h]	--	--	

BEE = butoxy ethanol ester; PGBE = propylene glycol butyl ether esters.

B) INSECTICIDES AND RELATED MATERIALS

Compound	Vapor pressure	MW				
DDT	3×10^{-7}[i]	354.5	0.0012[h]	370,000[c]	13.3	131,000 to 355,000
DDD	--	320	--	--	8.1	
DDE	--	318	--	--	24.0	
dieldrin	1.4×10^{-6}[i]	381	0.10[i] 0.25[h]	--	10.1	

endrin	2×10^{-7}	381	0.23^h	--	11.5
aldrin	4.9×10^{-5i}	365.5	0.01^i	--	32.3
methoxychlor	--	345.5	0.1^h	--	3.6
p-dichlorobenzene	1^c	147	80^h	$2,399^j$	--
diphenyl ether	$(20°C)$ 137^c	170	21^c	$15,850^j$	--
2-biphenyl ether	--	246	--	$457,100^j$	--
hexachlorobenzene	--	285	insol.^{h}	$1,660,000^j$	--
PCB 2,2',4,4'-tetra-chlorodiphenyl	--	292	--	$5,248,000^j$	--
chlorpyrifos	1.87×10^{-5}	350.5	0.4	$66,000^c$	--

C) CHLORINATED ALIPHATIC HYDROCARBON FUMIGANTS AND SOLVENTS

Compound	Vapor Pressure 25°C (mm Hg)	Mol. Wt.	Sol. in Water 25°C ppm	n-Octanol: Water Partition Coefficient	Water:Air Partition Coefficient[k]
methylene chloride	426[k]	84.9	19,800[k]	--	10
chloroform	200[k]	119.4	7,950[k]	--	6.3
1,1,1-trichloroethane	123[k]	133.4	1,300[k]	--	1.5
trichloroethylene	74[k]	131.4	1,100[k]	--	2.1
tetrachloroethylene	19[k]	165.9	400[k]	182[l]	2.4
carbon tetrachloride	115[l]	154	800	436.5[l]	--

a Beroza et al., 1969.
b Hamaker and Thompson, 1972.
c Dow, 1974.
d Hamaker, 1972.
e Flint et al., 1974.
f Skelly and Lamparski, 1974.
g Bailey and White, 1970.
h Gunther, 1969--calculated for p value.
i Lichtenstein et al., 1970.
j Neely et al., 1974.
k Dilling et al., 1973 (inverse ratios).
l Calculated according to method of Hansch et al., 1972.

n-octanol:water coefficients for 2,4,5-T (which would apply to other acids, depending on their pKa) is that in buffered (Clark and Lubs, 1949) solutions the coefficient number increases from 0.17 and 0.53 to 7.1 at pHs 9, 7, and 5, respectively (approaching the pKa of 2.88 for 2,4,5-T in an ionic solution of 0.03%). A dramatic change in the partitioning coefficient occurs when the acid is esterified as per the butyl ester which then increases the partitioning coefficient of 2,4,5-T greatly to 64,000. This partitioning coefficient helps explain the great difference in apparent toxicity to fish of 2,4-D and the butyl ester of 2,4-D (Hughes and Davis, 1963; Davis and Hughes, 1963). Since the partitioning coefficients are greatly different, it is reasonable from this and the residue work of Rodgers and Stalling (1972) to assume that more 2,4-D is taken up by the fish as the ester then as the acid, thus subjecting the fish to a larger dosage (mg/kg basis) than that apparent from the concentrations applied to water in which the fish were exposed.

The n-octanol water binary partition coefficient values for pesticides indicates what might happen in solvents of greatly different solubilities. Since nature often has other impurities in water, partition coefficients obtained in tertiary systems such as equal volumes of (isooctane):(80% acetone 20% water) results in much smaller numbers than octanol:water partition coefficients. These partition coefficients vary from 0.01 to 32.3, aldrin being the highest value, while DDT was also high. The lower coefficient values compared to octanol:water reflect a more varied choice and volume of partitioning solvent. Only a few values have been calculated for soil organic matter:water partitioning coefficient, but DDT still has a value of up to 355,000. The air:water coefficients are discussed in greater detail in the section on pesticide volatility from water.

DISCUSSION AND CONCLUSIONS

Pesticides, just like other chemicals are taken up and distributed in gas, liquid, and solid phases of the environment. They distribute by flow systems (mainly water or air) to surfaces to which they adsorb to and are often absorbed by living organisms and return changed or unchanged to flow systems inside or outside of organisms. Three important environmental aqueous flow systems are bodies of water, blood, and plant sap which, while variable in flow rate, have the ability to rapidly distribute solutes and particulates to adsorptive surfaces, organisms, organs, and/or cells. These three transport systems are composed percentage-wise mainly of water, however, they contain small amounts of many chemicals with similar functional groups in common which modify the dominance of water. Because of these distribution systems, partitioning of pesticides in and from flow systems can be expected to occur onto surfaces according to inherent physical and chemical properties of the pesticide

such as volatility, water and fat solubility, and sorption to solids
in the presence of flow systems. Plant and animal organisms contain
and release water, food, and pesticides by partitioning them through
surfaces, organs, and cell barriers.

Pesticide volatility from water appears to increase with increasing amounts of water evaporation, and decreasing water solubility. Thus, even DDT is relatively quite volatile from water,
however, this may be influenced by cosolvents (such as soil organics)
or competitive adsorptive surface of soils in water such as soil
particulates or plants or animals.

Sorption of most pesticides, while variable on the mineral portion of soil, is mainly correlated with the soil organic matter content which has ingredients having sorptive capacity for both
ionized and un-ionized, and fat and water soluble pesticides. The
effect of the pH of soil on sorption of acidic and basic compounds
may also be important. Sorption on soil and organisms takes place
most rapidly under the greatest concentration differential between
the flow system and sorptive surface. Typically well over 50% of
the total adsorptive uptake of pesticides occurs in the first few
hours of exposure. While the total amount of pesticide adsorbed
on carbon and algae increases with increasing concentration in the
flow system (water) the concentration factor may decrease with increasing pesticide concentration in water. Clearance of pesticides
from tissues is not necessarily at the same rate as uptake, and is
often slower. Other factors affecting adsorption are temperature,
speed of circulation of the flow system, ratio of biomass to flow
system volume, and mode of intake of pesticide into the organism.

The distribution of DDT and other stable chemicals in fat of
animal organisms, if given sufficient time for equilibration, appears
to roughly correlate with the n-octanol:water partition coefficient
obtained as a guide to the bioconcentration factor in fat tissues
of animals from aqueous sources. This factor alone can be somewhat
difficult to correlate with fat under some conditions since, in fish,
fat is variously distributed in tissues at different ages, stages
of sexual development, and seasons of the year, sometimes being
rather uniformly distributed and other times not. Also some fat
tissues appear to be more rapidly metabolized than others and therefore releases pesticides into the blood at various times of stress
or according to the need for the fat. The greater the number and
volume of components in an ecosystem, the greater the complexity
of the distribution of the available pesticide from water, and less
likely the expectation of extremes in bioaccumulation. The distribution of a water and organic soluble compound like dalapon as indicated by the n-octanol:water partition coefficient, is in great
contrast to DDT, since it does not volatilize at a rate significantly different from water, nor partition in favor of animal organisms, or their fat, but does accumulate in some plant organisms.

Herbicides like 2,4-D and 2,4,5-T partition somewhat similarly to dalapon, but when made into oil soluble (the butyl ester) forms, the octanol:water partition coefficient is considerably higher. However, this is not very significant in nature since the ester bond is not very stable in animal organisms and thus the ester does not continue to bioaccumulate like DDT. Every chemical has a uniqueness in its combination of physical and chemical properties, and in the way in which it will act in the environment.

From pesticide bioaccumulation data it is apparent that the first period of residue pickup is due to adsorption and is often related to high surface area to mass ratio of the adsorbent and may result in high residues. With many pesticides no further significant uptake of residue after the first day is made during a continued exposure period of days or weeks. However, if the pesticide is high in fat solubility, low in water solubility and persistent in the environment, a continued increase in residues may take place over time until a steady state (equilibrium) point is reached in live organisms. Such residue pickup is by absorption and such bioaccumulating chemicals can be detected at an early stage by obtaining early baseline residue levels for comparison with later levels. Other very useful partition coefficients to obtain in addition to n-octanol to water are soil organic matter to water, and water to air. These partition coefficients can be calculated roughly from predetermined constants of water and n-octanol solubility and vapor pressure. These calculated coefficients are useful in determining the potential of pesticides for leaching through soil, and uptake and distribution throughout the biota. Distribution coefficients of soil organic matter to water, and of aquatic animal organisms to water are often of the same order of magnitude.

SUMMARY

Natural bodies of water, blood, and plant sap were compared as aqueous flow systems for transporting pesticides and other chemicals. The uptake and distribution of pesticides in biological systems were related to water solubility; volatility from water; adsorption to soil organic matter, plant and animal organisms in aquatic ecosystems; effects of temperature; type of flow system; route of pesticide intake; specific types of organism tissue; and ratio of biomass to surface area for sorption. Differences between bioadsorption and bioabsorption were discussed. Partition coefficients between n-octanol and water; soil and water; and water and air are important guides to the bioaccumulation potential of pesticides.

ACKNOWLEDGMENT

The help given by Dean R. Branson, Wendell L. Dilling, Cleve A. I. Goring, George E. Lynn, Richard W. Meikle, Grant N. Smith, and Theodore R. Torkelson of The Dow Chemical Company in making this a better publication is greatly appreciated.

REFERENCES

Ahlrichs, J.L. 1972. The soil environment. In, Organic Chemicals
in the Soil Environment. C.A.I. Goring and J.W. Hamaker,
editors. Marcel Dekker, Inc., New York.

Altman, P.L. 1961. Blood and other body fluids. Biological
Handbook. Fed. Amer. Soc. for Exper. Biology. Washington,
D.C.

Bailey, G.W., and J.L. White. 1970. Factors influencing the adsorption, desorption, and movement of pesticides in soil. Res.
Rev. 32:29.

Beroza, M., M.N. Inscoe, and M.C. Bowman. 1969. Distribution of
pesticides in immiscible binary solvent systems for cleanup
and identification and its application in the extraction of
pesticides from milk. Res. Rev. 30:1.

Branson, D.R., G.E. Blau, H.C. Alexander, D.R. Thielen, and W.B.
Neely. In press. Steady-state bioconcentration methodology
for 2,2',4,4'-tetrachlorobiphenyl in trout. Trans. Am. Fish.
Soc.

Cope, O.B., E.M. Wood, and G.H. Wallen. 1970. Some chronic effects
of 2,4-D on the Bluegill (*Lepomis machrochirus*). Trans. Amer.
Fish Soc. 99(1):1.

Cope, O.B., J.P. McCraren, and L. Eller. 1969. Effects of dichlo-
robenil on two fishpond environments. Weed Science 17(2):158.

Cox, J.L. 1971. Uptake, assimilation, and loss of DDT residues by
Euphausia pacifica, a euphausid shrimp. Fishery Bulletin 69
(3):627.

Crosby, D.G. 1972. The photodecomposition of pesticides in water.
Fate of organic pesticides in the aquatic environment. Adv.
Chem. Ser. 111, American Chemical Society, Washington, DC,
p. 173.

Crosby, D.G., and R.K. Tucker. 1971. Accumulation of DDT by
Daphnia magna. Environ. Sci. Technol. 5:714.

Davis, J.C. 1970. Estimation of circulation time in the Rainbow
trout, *Salmo gairdneri*. J. Fish. Res. Board Canada 27(10):1860.

Davis, J.T., and J.S. Hughes. 1963. Further observations on the
toxicity of commercial herbicides to Bluegill Sunfish. Proc.
So. Weed Conf. 16:337.

Dilling, W.L., N.B. Tefertiller, and G.J. Kallos. 1973. Environ-
    mental fate of chlorinated solvent. I. Evaporation rates and
    reactivities of methylene chloride, chloroform, 1,1,1-trichloro-
    ethane, trichloroethylene, tetrachloroethylene, and other
    chlorinated hydrocarbons in dilute aqueous solutions. The Dow
    Chemical Company. 30th Annual Fall Scientific Meeting, Midland,
    Michigan.

Dow Chemical Company. 1974. Unpublished data.

Flint, G.W., J.J. Alexander, and O.P. Funderbuck. 1968. Vapor
    pressures of low-volatile esters of 2,4-D. Weed Science 16(1):
    541.

Goring, C.A.I. 1967. Physical aspects of soil in relation to the
    action of soil fungicides. Annual Review of Phytopathology 5:
    285.

Goring, C.A.I. 1972. Agricultural chemicals in the environment,
    a quantitative viewpoint. In, Organic Chemicals in the Soil
    Environment. C.A.I. Goring and J.W. Hamaker, editors. Marcel
    Dekker, Inc., New York.

Goring, C.A.I., and J.W. Hamaker. 1972. Organic Chemicals in the
    Soil Environment. Volumes I and II. Marcell Dekker, Inc.,
    New York.

Grzenda, A.R., D.F. Paris, and W.J. Taylor. 1970. The uptake,
    metabolism, and elimination of chlorinated residues by goldfish
    (*Carassius auratus*) fed a ^{14}C-DDT contaminated diet. Trans.
    Amer. Fish. Soc. 2:385.

Gunther, R., W.E. Westlake, and P.S. Jaglan. 1968. Reported solu-
    bilities of 738 pesticides in water. Res. Rev. 20.

Hamaker, J.W. 1972. Diffusion and volatilization. In, Organic
    Chemicals in the Soil Environment. C.A.I. Goring and J.W.
    Hamaker, editors. Marcel Dekker, Inc., New York.

Hamaker, J.W., and J.M. Thompson. 1972. Adsorption. In, Organic
    Chemicals in the Soil Environment. C.A.I. Goring and J.W.
    Hamaker, editors. Marcel Dekker, Inc., New York.

Hansch, C., A. Leo, and D. Nikaitani. 1972. On the additivity-
    constitutive character of partition coefficients. J. Org.
    Chem. 37:3090.

Hardy, J.L. 1966. Effect of TORDON[R] herbicides on aquatic chain
    organisms. In Down to Earth. p. 11.

Hewitt, E.A. 1948. Diseases of Poultry. H.E. Biester and L.N. Schwarte, editors. Second Edition. Iowa State University, Ames, Iowa.

Hughes, J.S., and J.T. Davis. 1963. Variations in toxicity to Bluegill Sunfish of phenoxy herbicides. Weeds 11(1):50.

Isensee, A.R., P.C. Kearney, E.A. Woolson, G.E. Jones, and V.P. Williams. 1973. Distribution of alkyl arsenicals in model ecosystem. Environ. Sci. Technol. 7(9):841.

Johnson, B.T., and J.O. Kennedy. 1973. Biomagnification of p,p'-DDT and methoxychlor by bacteria. Applied Microbiology 26(1): 66.

Johnson, B.T., C.R. Sanders, and H.O. Sanders. 1971. Biological magnification and degradation of DDT and aldrin by freshwater invertebrates. J. Fish. Res. Board of Canada 28(5):705.

Kapoor, I.P., R.L. Metcalf, R.F. Nystrom, and G.K. Sangha. 1970. Comparative metabolism of methoxychlor, methiochlor, and DDT in mouse, insects, and in a model ecosystem. J. Agr. Food Chem. 18(6):1145.

Kenaga, E.E. 1956. An evaluation of the use of sulfur dioxide in fumigant mixtures for grain treatment. J. Econ. Entomol. 49(6): 723.

Kenaga, E.E., W.K. Whitney, J.L. Hardy, and A.E. Doty. 1965. Laboratory tests with dursban insecticide. J. Econ. Entomol. 58(6):1043.

Kenaga, E.E. 1972. Guidelines for environmental study of pesticides: determination of bioconcentration potential. Res. Rev. 44:73.

Kramer, P.J., and T.T. Kozlowski. 1960. Physiology of Trees. McGraw-Hill Book Company, Inc., New York.

Lang, A. 1949. Clark and Lubs Acid-Base Indicators. Handbook of Chemistry. Handbook Publishers, Inc., Sandusky, Ohio. p. 11.

Leo, A.J. 1972. Relationships between partitioning solvent systems. Adv. Chem. Ser. 114. American Chemical Society, Washington, DC.

Leo, A.J., C. Hansch, and D. Elkins. 1971. Partition coefficients and their uses. Chem. Rev. 71:525.

Lichenstein, E.P., and K.R. Shultz. 1970. Volatilization of insecticides from various substrates. J. Agr. Food Chem. 18(5): 814.

Lincer, J.L., J.M. Dolon, and J.H. Nair, III. 1970. DDT and endrin fish toxicity under static versus dynamic bioassay conditions. Trans. Amer. Fish. Soc. 99(1):13.

MacKay, D., and A.W. Wolkoff. 1973. Rate of evaporation of low-solubility contaminants from water bodies to atmosphere. Env. Sci. Tech. 7(7):611.

McCraren, J.P., O.B. Cope, and L. Eller. 1969. Some chronic effects of diuron on Bluegills. Weed Science 17(4):497.

Meikle, R.W., N.H. Kurihara, and C.R. Youngson. 1972. A bioaccumulation study utilizing DDT and a mosquito fish, *Gambusia* sp., in an aquatic environment. Unpublished report. Dow Chemical U.S.A., Walnut Creek, California.

Metcalf, R.L., G.K. Sangha, and I.P. Kapoor. 1971. Model ecosystem for the evaluation of pesticide biodegradability and ecological magnification. Env. Sci. and Tech. 5(8):709.

Neely, W. Br., D.R. Branson, and G.E. Blau. 1974. Predicting the bioconcentration potential of organic chemicals in fish. Unpublished Report. The Dow Chemical Company, Midland, Michigan.

Premdas, F.H., and J.M. Anderson. 1962. The uptake and detoxification of C^{14} labeled DDT in Atlantic salmon, *Salmo solar*. J. Fish. Res. Board Canada 30:837.

Reinert, R.E. 1970. Pesticide concentrations in Great Lakes fish. Pest. Mont J. 3(4):233.

Reinert, R.E. 1972. Accumulation of dieldrin in an algae (*Scenedesmus obliquus*), *Daphnia magna*, and the guppy (*Poecilia reticulata*). J. Fish. Res. Board Canada 29(10):1413.

Reinert, R.E., and H.L. Bergman. 1974. Residues of DDT in lake trout (*Salvelinus namaycush*) and coho salmon (*Oncorhynchus isutch*) from the Great Lakes. J. Fish. Res. Board Canada.

Reinert, R.E., L.J. Stone, and W.A. Willford. 1974. Effect of temperature on accumulation of methylmercuric chloride and p,p'-DDT by Rainbow Trout (*Salmo gairdneri*). J. Fish. Res. Board Canada.

Rodgers, C.A., and D.L. Stalling. 1972. Dynamics of an ester of 2,4-D in organs of three fish species. Weed Science 20(1):101.

Sanders, H.O., and J.H. Chandler. 1972. Biological magnification of a polychlorinated biphenyl (Aroclor[R] 1254) from water by aquatic invertebrates. Bull. Env. Cont. Toxi. 7(5):257.

Skelly, N.E., and L.L. Lamparski. 1974. Laboratory-simulated environmental study for 2,4,5-T acid and its butyl ester-partition coefficients. Unpublished Report. The Dow Chemical Company, Midland, Michigan.

Smith, G.E., and B.G. Isom. 1967. Investigation of effects of large-scale applications of 2,4-D on aquatic fauna and water quality. Pest. Mon. J. 1(3):16.

Smith, G.N., B.S. Watson, and F.S. Fischer. 1966. The metabolism of ^{14}C 0,0-dimethyl 0-3,5,6-trichloro-2-pyridyl phosphorothioate in fish. J. Econ. Entomol. 59(6):1464.

Smith, G.N., F.S. Fischer, and R.J. Axelson. 1970. The volatilization and photodecomposition of Plictran[R] miticide. Unpublished Report. The Dow Chemical Company, Midland, Michigan.

Smith, G.N., Y. Taylor, and B.S. Watson. 1972. Ecological studies on chlorpyrifos. Unpublished Report. Chemical Biology Research, The Dow Chemical Company, Midland, Michigan.

Smith, G.N., Y.S. Taylor, and B.S. Watson. 1972a. Ecological studies on dalapon (2,2-dichloropropionic acid). Unpublished Report. The Dow Chemical Company, Midland, Michigan.

Smith, G.N., Y. Taylor, and B.S. Watson. 1972b. Ecological studies on chlorpyrifos. Unpublished Report. The Dow Chemical Company, Midland, Michigan.

Spector, W.S. 1956. Handbook of Biological Data. W.S. Saunders Company, Philadelphia.

Stevenson, F.J. 1972. Organic matter reactions involving herbicides in soil. J. Environ. Quality 1(4):333.

Sturkie, P.D. 1965. Avian Physiology. Cornell University Press. Second Edition, Ithaca, New York.

Torkelson, T.R., R.M. Kary, E.A. Pfitzer, W.E. Reinhart, and T.F. Hatch. The uptake storage and elimination of inhaled 1,1,1-trichloroethane in the rat. University Microfilm 67-3045. Ann Arbor, Michigan.

Walker, C.R. 1971. Chemicals and their effects on our aquatic
    environment. Presented at the 24th Annual Meeting of the SE
    Weed Science Society, Memphis, Tennessee.

Weber, J.B. 1972. Interaction of organic pesticides with parti-
    cular matter in aquatic and soil systems. Fate of organic
    pesticides in the aquatic environment. Adv. Chem. Ser. 111.
    American Chemical Society, Washington, DC.

Weber, W.J., Jr., and J.P. Gould. 1966. Sorption of organic
    pesticides from aqueous solution. Adv. Chem. Ser. 60. Ameri-
    can Chemical Society, Washington, DC.

Wershaw, R.L., P.J. Bucar, and M.C. Goldberg. 1972. Interaction
    of pesticides with natural organic material. Environ. Sci.
    Technol. 3:271.

Whitney, W.K., and E.E. Kenaga. 1960. Distribution and sorption
    of liquid fumigants applied to wheat by recirculation. J.
    Econ. Entomol. 53(2):259.

Whitney, W.K., and E.E. Kenaga. 1966. Bioassay experiments to
    determine the volatility of tricyclohexyltin hydroxide. Un-
    published Report. The Dow Chemical Company, Midland, Michigan.

Youngson, C.R., and R.W. Meikle. 1972. The effect of water tem-
    perature on the bioaccumulation of ^{14}C-DDT by mosquito fish,
    Gambusia sp. Unpublished Report. Dow Chemical U.S.A., Walnut
    Creek, alifornia.

INTERNATIONAL DYNAMICS OF PESTICIDE POISONING

J.E. Davies and S.A. Poznanski
Department of Epidemiology and Public Health
University of Miami School of Medicine
Miami, Florida 33152

R.F. Smith
Department of Entomological Sciences
University of California
Berkeley, California 94720

V.H. Freed
Department of Agricultural Chemistry
Environmental Health Sciences Center
Oregon State University
Corvallis, Oregon 97331

INTRODUCTION

With present population trends, increased food and 'fiber production is one of the central problems for all nations. In the
past, agricultural chemicals have been the mainstay of all expanding
food production programs and their continued use will be one essential response to these increased demands. The growing problem of
pest resistance and environmental degradation, however, indicated
that their future use will have to be part of a broad integrated
pest control program utilizing a wide variety of modern pest management techniques of which the use of chemicals will be one component. No longer can reliance be placed solely on chemical control.
The future role of pesticides will have to be more judicious and
restrained, respecting the benefits of natural predators in the
role of controlling harmful pests.

Pesticide usages and pest management practices like everything
else have undergone profound and rapid changes in the last decade.
Problem of pest resistance, and primary and secondary pest resur-
gence, increased production and registration costs, increased

human toxicity problems, environmental degradation, changing agricultural and social patterns are some of the factors which are
responsible for these changes. Recent pest management surveys
conducted in widely different geographic areas by multidisciplinary
teams by the University of California/U.S. AID project and others
have reported that in many areas of the world, the use of agricultural pesticides is poorly understood and in certain areas impru-
dent and excessive (Koehler et al., 1972; Apple and Smith, 1972;
Cavin et al., 1972; Echandi et al., 1972; Sasser et al., 1972; and
Glass et al., 1972). Serious problems, some of them of far reaching
consequences have been encountered and have resulted, on occasion,
in appeals to the different international organizations for assistance and special training in pesticide management. Assistance has
been sought also for integrated pest control guidance advice on
pesticide legislation and registration. The most areas of concern
were: 1) increases of human pesticide intoxication; 2) residue
contamination of food with pesticide levels being too high for
international trade; and 3) malarial resurges primarily due to
vector resistance acquired in part as a result of agricultural pesticide use.

During a recent visit by project representatives of the U.S.
AID project to El Salvador, examples of all these problems were
encountered. This report describes the findings of these surveys
and reports on an ingenious organizational framework developed by
Health and Agricultural representatives of El Salvador during a
recent pesticide management training organized by U.S. AID to overcome these problems.

PESTICIDE MANAGEMENT PROBLEMS AND THEIR RESOLUTION IN EL SALVADOR

El Salvador is situated on the Pacific coastal plain of Central
America and covers an area of 21,000 square kilometers; the population is approximately 3.8 million people. Much of the region is
hilly and mountainous, and cotton is the major agricultural commodity; this product is largely confined to the valleys and coastal
plains. Substantial amounts of pesticides are being used for agricultural products and in public health for malarial control. It
is estimated that 8,000,000 pounds of pesticides are imported
annually, 60% of which is ethyl and methyl parathion. Most of this
is used on cotton, often with 18 to 20 applications to a crop. A
mixture of DDT and toxaphene is also used. In this area, there were
numerous reports of pesticide poisonings. Although human poisonings
is not a notifiable disease, excellent records of hospitalized
cases of pesticide poisonings from 1964-1972 had been kept by the
public health authorities in the Ministry of Health (Valverde and
Ruegas, 1973), and these are as follows:

Year	Non-Fatal Cases	Fatal Cases
1963	1,104	11
1964	965	2
1965	938	1
1969	584	7
1970	474	7
1971	586	10
1972	2,787	5

The incidence data for 1972 are important not only because they identified a greater than four-fold increase over the preceding year, but also because the 1972 total was approximately reconfirmed by another survey conducted by representatives from the Ministry of Labor. These investigators made a survey from all areas of the country obtaining hospital and non-hospital incidence data of pesticide poisoning by questioning. Their survey identified 2,028 pesticide poisoning cases; these included 202 women, 224 children, and there were 25 male fatalities, 3 children fatalities, and 2 female fatalities. Alvarenga (1973) reported that in 1972, 942 cases were treated at the hospital; 747 cases of organophosphate intoxications were admitted and 195 of chlorinated hydrocarbon intoxication cases. The authorities felt that there were several factors contributing to the increase in 1972. Among these was the possibility of a re-entry problem. Due to the persistence of dislodgeable foliar pesticide residues in dry areas of the world, the time interval between pesticide application and worker re-entry has been shown to be a critical variable in the poisoning of pickers (Milby et al., 1964; Gunther, 1969). In El Salvador, agricultural officials felt that the original period of 72 hours which had been used as the minimal time before pickers could enter the fields following the early application of parathion had been too short, and was partially instrumental in the occurrence of the increased incidence of these cases. The first use of ultra-low volume application of parathion was thought to be another factor, and the unusually dry season with little rainfall to wash off the pesticide residue considered to be the third casual factor. Many of the dwellings of the workers were located on the edge of the cotton fields and human exposure through overspray was inevitable.

PESTICIDE POISONINGS IN OTHER AREAS

Surveys in other parts of Central America have also indicated that human pesticide poisoning is a serious problem. Here, cotton is the major crop and ethyl and methyl parathion mixtures have been widely used. There are several reports of organophosphate poisoning statistics: Vandekar (1972), published the following statistics from Guatemala and Nicaragua:

Year	Guatemala	Nicaragua
1968	1,374	
1969	837	258
1970	659	221
1971	1,100	356

From a global point of view, reporting of widespread poisoning in pesticide workers and in the general population are numerous. Although accurate statistics are not available it is obvious that the problem is both extensive and serious. The World Health Organization estimates that there are approximately 500,000 cases occurring annually with a greater than 1% mortality rate (WHO, 1973).

Pesticide poisonings were also recognized during a survey of health problems for the World Health Organization in Indonesia in 1969; at this time, the organochlorines were the major offenders. The data although incomplete was acquired during a visit to different centers in Java and Sumatra revealed that 478 organochlorine poisonings and 31 organophosphate poisonings had occurred, and there were reports of 125 deaths (Davies, 1972). Most of the data listed only poisoning incidents which involved more than one person. The most common mechanism stemmed from the accidental contamination of food from discarded pesticide containers. There were numerous reports of groups of people who became ill, vomited, and had convulsive seizures following ingestion of food which had been prepared in a pesticide container. The diagnosis was usually made on epidemiological grounds and was seldom confirmed by analytical toxicological studies.

The inherent toxicity of the pesticide being used appears to be one of the most important variables, and more and more reports of human intoxication will occur if their introduction is not preceded by a vigorous pesticide management training program. The problem is especially acute in warmer parts of the world, and the problem is not unique to the United States. For example, when the organophosphates were first widely used in the United States, similar pesticide problems were encountered especially in Florida.

Nine years ago when all possible human health effects from pesticide usage were being studied through the Community Studies Pesticides Program, there were initially flagrant examples of misuse and pesticide poisoning was a common emergency in Dade County, Florida. Itinerant sales were peddling parathion in unlabelled brown paper bags and selling the product as a "roach killer"; parathion could be bought in the local stores and even the local barber shops, headlines of pesticide poisoning was a common occurrence in the local papers. Cases were occurring in adults and children alike and were especially frequent among the black population (Davies et al., 1965; Reich et al., 1968; Davies et al., 1970). The main reason for this was because parathion was readily accessible to this population group and was frequently brought into the urban setting from the fields; discarded pesticide containers were used for a variety of purposes and often contained small amounts of pesticide residual (Davies et al., 1973). The inherent toxicity of parathion was ill understood and cases occurred as a result of accidents, worker exposure, and in suicidal and homicidal attempts. Publicizing the dangers of this agricultural chemical in the mass media was another important factor in promoting the danger of pesticides. Table 1 lists the fatalities in Dade County and the State of Florida (Bureau of Vital Statistics, 1973) from 1965 to 1972. The clinical recognition and management of this medical emergency has also improved contributing to the declining mortality rate. Organophosphate poisoning can mimic epilepsy, pneumonia, diabetes, and many other conditions. In the early days, the condition was not readily identified and early diagnosis is an essential contributant for a favorable outcome. The case fatality rate for parathion poisoning in Dade County, a rate which expresses the chances for dying from a particular condition, and which, is in part a measure of the efficacy of medical treatment, this declined from 26% to 0% in 1972 in Dade County. These data confirm the improved clinical management of this condition over the years. Similar improvements can be expected in those areas of the world now going through an epidemic of pesticide poisonings; legislation, education, and training of the occupational workers must be the key to the problem.

PERSISTENCE

Persistence and the consequences of food and human contamination was a second problem of pesticide management which was readily appreciated in El Salvador. In addition to the cotton production,

TABLE 1

Parathion Poisoning Cases in Dade County, Florida, and
Pesticide Mortality and Morbidity in the State
of Florida, 1965-1972

	Dade County		Florida	
Year	Fatalities	Non-Fatalities	Mortality	Morbidity
1965	5	14	22	649
1966	6	7	22	N/A[a]
1967	4	6	35	437
1968	2	5	26	484
1969	3	3	17	365
1970	2	7	13	299
1971	0	5	8	323
1972	0	5	7[b]	N/A

[a] Data not available.
[b] Incomplete data.

a sizable meat industry has developed and the main grazing pasture
is on the coastal plains in areas which adjoin the areas of cotton
production. In addition to ethyl and methyl parathion, a DDT-
toxaphene mixture was being used on cotton to contain pests, and
cottonseed oil was used for fattening beef. Corn, too, was used
for fattening purposes, and aldrin was the main pesticide used to
control pests. It was being used as a pre-plant treatment and for
treatment and for treatment of the crop after emergence. There was,
therefore, ample opportunity for contamination of agricultural beef
stock. Reports of excessive pesticide residues in beef cattle were
numerous and on several occasions exportation of beef lots to the
United States had been prohibited on account of excessive pesticide
residues, restrictions wherein posed a serious threat to the export
industry and the general economy of the country.

Although these contaminations were of serious import to the cattlemen and to agricultural officials, concern for the residue problem was also expressed by representatives of the Ministry of Health. Health officials were concerned that food, designated for export but which failed to meet international standards, might be redirected for human consumption, and they were especially concerned for human contamination in addition to the animal contamination. They regretted the absence of baseline information on general population levels of persistent pesticides, and expressed the hope that future residue monitoring programs of food and feed stock would also include a subsample of fat analyses obtained from persons undergoing surgery.

RESISTANCE

Resistance was the third pesticide management problem encountered in El Salvador. Malaria was uncompletely controlled and Nave-Rebollo (1971) reported that "the malarial situation of the country at the end of 1970 had not improved substantially and in fact the situation had deteriorated. Malariologists were concerned with the appearance of mosquito resistance to the carbamate insecticide (propoxur) and DDT. The greatest incidence of malaria in El Salvador is along the coastal plains. The carbamate resistance found in these areas almost certainly reflected cross resistance acquired from the application of organophosphates to cotton. Anopheles Albumanis, the malarial vector in El Salvador, in certain areas was resistant to DDT, dieldrin, and propoxur. As has been reported in other areas of the world, agricultural pesticide usage has interface and interfered with the public health control of malaria.

AN AGROMEDICAL SEMINAR

The recognition of these pesticide management problems prompted a pesticide managment training program in El Salvador in December, 1973. Developed by the U.S. AID project with the University of California, and with assistance from the Pan American Health Organization, the emphasis was on "agro-medical" approach, with in-country representatives formulating their own recommendations for solutions to the problems. The theme of the seminar was "Pesticide Management in El Salvador" with emphasis on the problems of persistence, resistance, and poisonings. Through a multidisciplinary "agro-medical" approach pesticide management as a preliminary step to the ultimate goal of integrated pest control was developed. The seminar was opened with incisive statements by the Minister of Health indicating the committment of their government to the program and outlining the needs for the seminar and setting forth the objectives. Six experts from the U.C./U.S. AID project,

six experts from El Salvador, and two from the Pan American Health
Organization made technical presentations. The topics ranged from
chemistry-toxicology and the chemodynamics of pesticides, integrated pest control and malaria, as they related to El Salvador.
In order to obtain active involvement of the country, technical
committees were selected by the floor and every one of the 88
participants joined one of the groups. The groups identified the
specific pesticide management problems of their country and then
sought to enunciate objectives and devise programs and methods of
implementation of such programs for the solution of those problems.
At the conclusion of the Seminar, a certificate was signed by the
two Ministries, U.S. AID and the U.C./U.S. AID Pest Management
Project was provided to each participant.

FOLLOW-UP TRAINING

Following the seminar, an intensive training program of chemists was implemented and conducted for two weeks. Time was spent
in the laboratory working with the chemists, reviewing instrumentation and introducing new analytical methodology for pesticide
analyses on a micro-scale. A chemist follow-up training program
has just been completed.

In addition to the chemists, provisions are being made in the
future for the intensive specific training of the agricultural and
public health disciplines. Preliminary steps were taken in conjunc-
tion with P.A.H.O. and the Ministry of Health in El Salvador to
identify a physician who had some skill in toxicology and epidemiology to be trained in the University of Miami School of Medicine,
and a senior agronomist to be trained in integrated pest control
technology at the University of California. A three-month training
program in agricultural chemistry will be provided by Oregon State
University.

CONFERENCE EVALUATION

Evaluation of any training seminar is not easy, and with a
topic as complex as pest management, progress can only be measured
some time after the initial training endeavor. This was certainly
the case for the ultimate evaluation of the program in El Salvador.
There were, however, certain immediate responses observed in the
training program which suggested that the initial impact of this
program was highly successful and augurs well for the future. The
first of these events was the number of attendants. As previously
mentioned, 88 individuals attended the entire course. The total
registration was as follows:

Ministry of Health	19 participants
Ministry of Labor	4 participants
Ministry of Agriculture	27 participants
University of San Salvador	8 participants
Industrial (in-country)	16 participants
Other out-of-country representatives: Speakers ICAITI, AID-Washington, and Project Personnel	14 participants

A second indication of the success of the program was the number and caliber of questions raised by the participants from the floor. Approximately 70 questions were raised from the floor and referred to a Technical Advisory Committee. These questions were indicative of the close scrutiny and grasp of the topics that the participants exhibited and were testimony to the interest of the groups.

Third and finally, the results of the workshop can also be measured by the recommendations and resolutions formed by each technical group. They were truly reflective of multidisciplinary agro-medical approach and the subject matter of the proposals ranged over a wide variety of topics. The organization in the near future of a central inter-agency pesticide commission was proposed with representation on it from the Ministers of Agriculture, Public Health, Labor, Economy, and Defense, and also the National University of El Salvador and private industry. This academic body which would be inaugurated by Executive Decree would legislate and coordinate the use of pesticides for the development of the country and the protection of the population and the environment. So constituted it would develop pesticide residue tolerances, develop legislation for the impartation of new pesticides and totally investigate pesticide management problems through the mechanism of regional pesticide protection teams. Joining and compliance with the requirements of the Codex Alimentarious was recommended, a step which would require continued upgrading of laboratory performance and the establishment of a quality control program. Collaboration with pesticide residue activities in other Central American countries and Panama, and participation with INCAP ICAITI was stressed. The toxicological proposals included the recommendation of a Chair of Toxicology in the medical school, and the training of members of the regional multidisciplinary pesticide protection teams. It was recommended that the latter should be capable of investigating,

TABLE 2

Agromedical Problems in Pesticide Management and Future Goals

Problems	Concerns		Agromedical Solution
	Health	Agriculture	
I. Pesticide poisoning	Epidemic human poisoning	Illness and time loss of worker	Investigation, prevention and surveillance
II. Persistance	Human contamination	Animal contamination and unacceptable food residues	Human and food residue monitoring programs
III. Resistance	Malaria resurgence	Pest control problems	Agromedical collaboration with integrated pest control programs

analyzing, and reporting human and environmental pesticide poisoning problems to the Central Commission. Ongoing blood and adipose surveys were recommended in man as well as food and animals.

The recommendations for training and education proposed the future education of chemists, public health personnel and agronomists who would be supported by an updated pesticide reference library.

The breadth of these new recommendations developed and approved by agronomists, public health people and industry as an evaluative measure of the success of the conference and all agreed that these recommendations would be presented to the appropriate ministers for their final consideration.

CONCLUSION

The three important pesticide management problems encountered in El Salvador were problems of equal import to both agriculture and public health. In unresolved, the consequences would be serious to both disciplines. The pesticide training program in El Salvador recommended the collaborative approach to their resolution and can be summarized in Table 2.

These recommendations which were proposed by in-country representatives of agriculture, clinical medicine, public health, and industry were submitted to the Ministers of Health and Agriculture. Their future implementation would go far to alleviate the present management problems in El Salvador.

ADDENDUM

In a follow-up visit by two members of the project team to El Salvador in June, 1974, the following additional information on the three pesticide management problems were obtained for 1973. The number of acute pesticide poisonings had dropped to 1,222 (Department of Epidemiology, Ministry of Health, El Salvador, 1974). In addition, it was learned that pesticide contamination of meat had much improved and no further rejection of beef exports on account of excessive pesticide residues had been experienced in 1973. On the other hand, malarial officials were greatly concerned by the continued increase in the number of positive cases of malaria with a highly significant increase in Plasmodium falciparium positive smears (Rebollo, 1974). In selected areas wherein Anopheles Albumanis was propoxur resistant chloroquine being used for malarial control purposes rather than controlling the vector by insecticides.

REFERENCES

Alvarenga, M.T.   1973.   Sintomas y tratamiento de envenenamiento a causa de plaguicidas. Management of Pesticides and Protection of the Environment Seminar, December 3-7.

Apple, J.L., R.F. Smith.   1972.   A preliminary study of crop protection problems in selected Latin American countries. University of California Technical Report.

Bureau of Vital Statistics.   1973.   Department of Health and Rehabilitative Services of Florida State.

Cavin, G.E., D. Raski, R.G. Grogan, and O.C. Burnside.   1972.   Crop protection in the Mediterranean basin. A multi-disciplinary team report. University of California Technical Report.

Davies, J.E.   1972.   Recognition and management of pesticide toxicity. In: Advances in Internal Medicine. G.H. Stollerman, editor. Yearbook Medical Publishers, Chicago.

Davies, J.E., J.C. Cassady, and A. Raffonelli.   1973.   The pesticide problems of the agricultural worker. In: Pesticides in the Environment: A Continuing Controversy. W.B. Deichman, editor. Symposia Specialists, Miami.

Davies, J.E., J.S. Jewett, J.O. Welke, A. Barquet, and J.J. Freal.   1970.   Epidemiology and chemical diagnosis of organophosphate poisoning. In: Pesticide Symposia. W.B. Deichman, editor. Halos and Associates, Miami.

Davies, J.E., J.O. Welke, and J.L. Radomski.   1965.   Epidemiological aspects of the use of pesticides in the south. J. Occup. Med. 7:12.

Echandi, E., J.K. Knoke, E.L. Nigh, Jr., M. Shenk, and G.T. Weekman.   1972.   Crop protection of Brazil, Uruguay, Bolivia, Ecuador, and Dominican Republic. University of California Technical Report.

Glass, E.H., R.J. Smith, I.J. Thomason, and H.D. Thurston.   1972.   Plant protection problems in Southeast Asia. University of California Technical Report.

Gunther, F.A.   1969.   Insecticide residues in California citrus fruits and products. Residue Reviews 28.

Koehler, C.S., R.D. Wilcoxson, W.F. Mai, and R.L. Zimdahl.   1972.   Plant protection in Turkey, Iran, Afghanistan, and Pakistan. University of California Technical Report.

Milby, T.H., F. Ottoboni, and H.W. Mitchell. 1964. Parathion residue poisoning among orchard pickers. JAMA 189:351.

Nave-Rebollo, O. 1971. Consejo Centroamericano de Salud Publica., 9th Annual Public Health Congress. San Salvador.

Rebollo, O.N. 1974. Personal Communication.

Reich, G.A., J.H. Davis, and J.E. Davies. 1968. Pesticide poisoning in South Florida. An analysis of mortality and morbidity and a comparison of sources of incidence data. Arch. Environ. Health 17:768.

Sasser, J.N., H.T. Reynolds, W.F. Meggitt, and T.T. Herbert. 1972. Crop protection in Senegal, Niger, Mali, Ghana, Nigeria, Kenya, Tanzania, and Ethiopia. University of California Technical Report.

Valverde and Ruegas. 1973. Personal communication. Ministry of Public Health, El Salvador.

Vandekar, M. 1972. Review of toxicity tests. Presented at Report on Seminar on Safe, Effective, and Utilization of Pesticides in Agriculture and Public Health in Central America and the Caribbean. Turrialba, Costa Rica, April 20-28.

Vandekar, M., R. Plestina, and K. Wilhelm. 1971. Toxicity of carbamates for mammals. Bull. WHO 44:241.

DETOXICATION OF PESTICIDES BY BIOTA

M.A.Q. Khan, M.L. Gassman, and S.H. Ashrafi

Department of Biological Sciences, University of Illinois

at Chicago Circle, Chicago, Illinois 60680

A pesticidal chemical, once released into the environment, is subject to physiochemical and biochemical processes which determine its fate and efficacy. The latter transformations involve biota and encompass biodegradation, detoxication, or, simply, metabolism. The metabolism of pesticides by living organisms has been the subject of various symposia (Institute fur Okologische Chemie, 1970; National Academy of Science, 1972; Khan and Hauge, 1970; Hodgson, 1969; Gillette et al., 1969; Matsumura et al., 1972; O'Brien and Yamamoto, 1970; American Chemical Society, 1973; Khan and Bederka, 1974). This article will present an overview of the knowledge of detoxication of pesticides by biota: microorganisms, plants, and animals.

Detoxication of a pesticidal chemical by biota may involve one or more of the following reactions: oxidations (aliphatic and aromatic), hydrolysis (ester and amide bonds), dealkylation (C-R or N-R bonds), dechlorinations, reductions, aromatic ring cleavage, etc. The detoxified molecule may then be excreted either as it is or following conjugation with a carbohydrate, amino acid, or lipid. A combination of both qualitative and quantitative differences in the detoxication reactions in biota is what determines the toxicity of a biocide to the interrelated members of ecosystems.

DETOXICATION REACTIONS

Oxidations

Pesticide oxidation may involve the introduction of phenolic or alcoholic hydroxyl groups into hydrocarbons; oxidation of alcohols to ketones or carboxylic acids; the oxidative removal of alkyl groups; formation of epoxides (this is not generally detoxication), N-oxides, sulfoxides, and sulfones; oxidative desulfuration of phosphorothionates (generally making the pesticide more toxic); and oxidative dehydrochlorination of photoisomers of cyclodienes (Table 1).

Aromatic hydroxylations. A benzene ring with Cl, alkyl, alkoxy, or NO_2 substituents is hydroxylated by removal of the substituent. Thus, phenoxy herbicides are hydroxylated in the *para*-position; e.g., 2,6-dichloro- and 2-chlorophenoxy-acetic acids (Thomas et al., 1964a). If this position is blocked (e.g., as in 2,4-D or 2,4,5-T), then the chlorine migrates to the 3- or 5-position (Thomas et al., 1964b). In the case of 2,4,5-T, however, some hydroxylation does occur at the 3-position (Figure 1). These reactions are common among plants and fungi (Thomas et al., 1964a; Thomas et al., 1964b; Faulkner and Woodcock, 1961). Phenols with chlorine in the *meta*- position are more resistant to biodegradation.

Biphenyl pesticides are hydroxylated by microorganisms to *cis*-diols via a cyclic peroxide and by multicellular organisms to *trans*-diols via an epoxide (Chapman, 1972). Carbaryl is hydroxylated at the 4- and 5-positions by fungi, plants, insects, and vertebrates (Liu and Bollag, 1971; Dorough and Casida, 1964; Eldefrawi and Hoskins, 1961). Vertebrates, in addition, hydroxylate it at the 5- and 6-positions (Leeling and Casida, 1966; Knaak et al., 1965). Carbaryl is the only insecticide known that is ring hydroxylated by microorganisms (Bollag and Liu, 1971; Still, 1968).

Aliphatic hydroxylations. Oxidation of alcohols and aldehydes is common among microorganisms. Even numbered aliphatic hydrocarbons undergo β-oxidation. An aliphatic acid undergoing β-oxidation requires two hydrogens on both the α- and the β-carbons. If one of these carbons has a substituent, then the substituent must be removed for oxidation (Chapman, 1972; Hegenman, 1972). This is why α-substituted phenoxy herbicides are not rapidly degraded.

In animals, the methyl group in toluene of *p*-nitrotoluene is rapidly oxidized to an alcohol or aldehyde. In higher alkylbenzenes, oxidation at the terminal carbon of the alkyl group produces a carboxylic acid while oxidation at other carbons produces secondary alcohols. However, the latter reaction involves only the penultimate and the α-methylenic position. α-Methylenic oxidation of DDT in insects produces kelthane (Figure 2) (Bowman et al., 1969; Perry, 1964).

O-Dealkylations. Oxidative removal of alkyl substituents is a common reaction in animals and plants; e.g., methoxychlor (Khan

et al., 1970a) (Table 1). Oxidative removal of methyl and ethyl
groups of phosphoric acid esters; e.g., malathion and parathion,
respectively (Figures 3 and 4), occurs less readily (O'Brien, 1967).
These pesticide reactions have not been clearly demonstrated in
microorganisms (Matsumura, 1974).

    N-Dealkylations. The efficiency of a plant to dealkylate sub-
stituted ureas and triazines (Figure 5) determines its tolerance
to these herbicides (Kearney et al., 1967; Bull, 1965; Alexander,
1972). A list of pesticides dealkylated by plants is shown in
Table 1.

    Insecticides, such as bidrin, undergo dealkylations in micro-
organisms, plants, and animals (Menzer and Casida, 1965; Bull and
Lindquist, 1964) (Figure 6). These reactions appear to increase the
toxicity of these insecticides (Menzer and Casida, 1965).

    In animals stable N-methylols can be produced from N-methyl-
carbamates and N-methylamides. Some insects and dogs can demethyl-
aminate alkylamines such as Zectran (Fawcett et al., 1954; Khan et
al., 1974). Microorganisms dealkylate alkylamines such as Zectran
and s-triazine (Matsumura, 1974; Benezet and Matsumura, unpublished
data; Kearney et al., 1967; Williams et al., 1972).

    Oxidative desulfurations. Desulfuration of phosphorothionates
occurs very readily in animals (Figures 3 and 4). Oxidative desulfuration is also common in plants (Bowman et al., 1969). These
reactions produce more toxic products but, in many cases, the de-
sulfurated product becomes more susceptible to degradation (Coppedge
et al., 1967). Microorganisms seem to be incapable of performing
these reactions (Ahmed and Casida, 1958, Williams et al., 1972).

    Oxidation of thioethers; e.g., disyston conversion to sulfo-
xides and sulfones, occurs in microorganisms, plants, and animals
(O'Brien, 1967; Bull, 1965; Coppedge et al., 1967). This can make
the pesticide more susceptible to biodegradation. For example,
Temik sulfoxide and sulfone (Figure 7) are readily hydroxylated by
the yeast, *Torulopsis*, and the alga, *Chlorella*, and various soil
microorganisms (Ahmed and Casida, 1958; Williams et al., 1972;
Coppedge et al., 1967). The soil microorganisms can, in addition,
convert Temik to the nitrile sulfoxide and oxime sulfoxide (Ahmed
and Casida, 1958; Williams et al., 1972; Coppedge et al., 1967).

Dechlorinations

    Since most of the first generation pesticides are chlorinated
hydrocarbons, their dechlorination constitutes another very important
series of detoxication reactions.

TABLE 1   Some examples of the oxidative reactions involved in the

Detoxication[a]	Microorganisms
Aryl hydroxylation	carbaryl (Bollag and Liu, 1971; Still, 1968)
Aliphatic hydroxylation	aldehydes, ketones (Thomas et al., 1964b; Faulkner and Woodcock, 1961; Chapman, 1972); β-oxidation of 2,4-dichlorophenoxyalkanoic acids (Chapman, 1972; Hegenman, 1972; Gutenmann et al., 1964).
O-dealkylation	- - - -
N-dealkylation	alkylamines, carbaryl (Ahmed and Casida, 1958); zectran (Benezet and Matsumura, unpublished data); simazine (Kearney et al., 1967)
N-methyl hydroxylation	- - - -
O-desulfuration	absent (Ahmed and Casida, 1958; Williams et al., 1972)
Sulfoxidation	Temik (Ahmed and Casida, 1958; Migamoto fenitrothion, sulfothion (Benezet and Matsumura, unpublished data)
Epoxidation	cyclodienes (Ahmed and Casida, 1958)
Epoxide hydration	dieldrin (Ahmed and Casida, 1958; Williams et al., 1972)

[a] Ether cleavage of 2,4-D in plants and ester cleavage of some phosphorothionates in animals are also oxidative reactions.

detoxication of pesticide by biota.

Pesticide Detoxified

Plants	Animals
carbaryl (Liu and Bollag, 1971; Dorough and Casida, 1964; Eldefrawi and Hoskins, 1961); 2,4-D (Thomas et al., 1964a).	carbaryl (Knaak et al., 1965); biphenyls (Clayson and Ashton, 1963); Baygon (Casida, 1969)
aldrin (Brooks, 1968); alkyl groups (Adamson, 1974; Kuhr, 1970); β-oxidation of MCPB (Wain, 1955; Fawcett et al., 1954)	cyclodienes (Khan et al., 1974); alkyl groups (Chakraborty and Smith, 1967; Smith, 1968); DDT (Bowman et al., 1969; Perry, 1964); p-nitrotoluene and other alkylbenzenes (Chakraborty and Smith, 1967); rotenone, pyrethrins (Casida, 1969)
Malathion, Abate (O'Brien, 1967; Blinn, 1968)	malathion, parathion (O'Brien, 1967; O'Brien, 1962); Baygon (Casida, 1969); methoxychlor (Khan et al., 1970a)
phenylureas, triazines (Nashed et al., 1970); diphenamid (Lemin, 1966); trifluralin (Matsunaka, 1972)	N-methyl- and N,N'-di-methyl carbamates, Bidrin (Menzer and Casida, 1965)
Bidrin, azodrin (Bull and Lindquist, 1964)	N-methyl carbamates (Casida, 1969); Bidrin (Menzer and Casida, 1965)
phosphorothioates and dithioates (O'Brien, 1967; Matsunaka, 1972; Miyamoto, 1972)	phosphorothionates (O'Brien, 1967)
Temik, Abate (Nashed et al., 1970; Miyamoto, 1972)	Temik (Andrews et al., 1967); Disyston (Bull, 1965)
cyclodienes (Yu et al., 1971)	cyclodienes (Khan et al., 1970a; Korte, 1968; Matthews and Matsumura, 1969; Ludwig and Korte, 1965)
dieldrin (Korte, 1968; Brooks et al., 1970)	cyclodienes (Korte, 1968; Matthews and Matsumura, 1969; Ludwig and Korte, 1965)

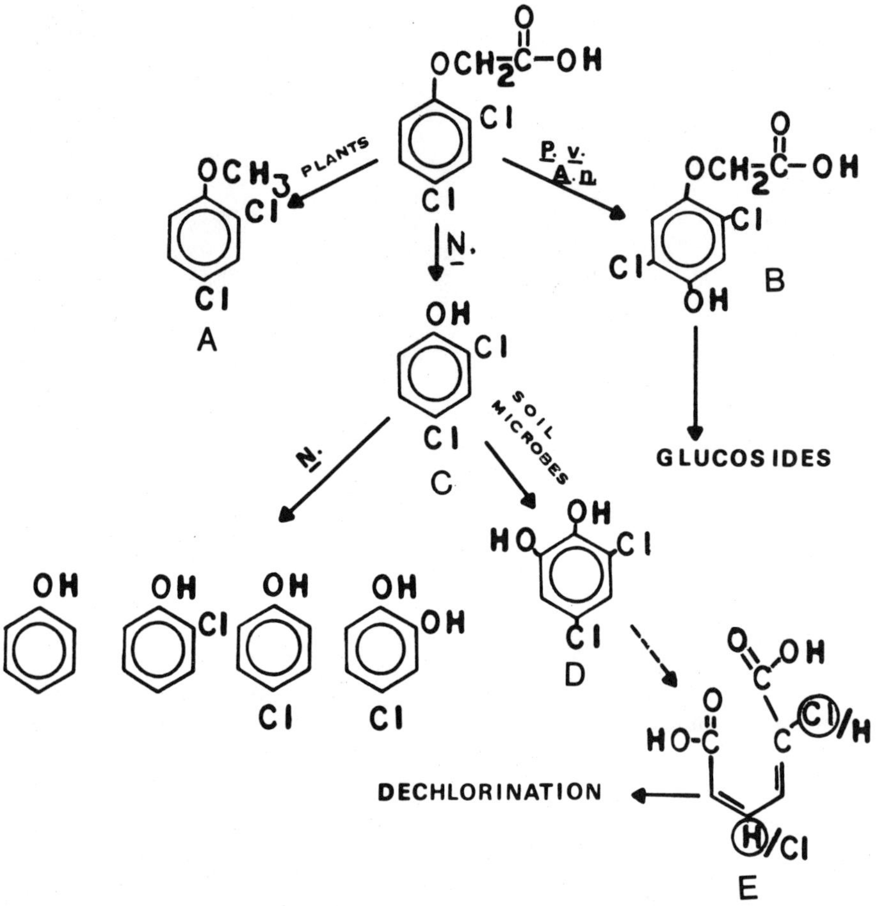

Figure 1.   Pathways of metabolism of the herbicide 2,4-dichlorophen-oxyacetic acid in biota.   Plants hydrolyze it to 2,4-dichloroanisole (A). *Pseudomonas vulgaris* and *Aspergillus niger* hydroxylate it at the 4-position with the chlorine migrating to the 5-position (B).   The latter can also shift the chlorine to the 3-position and form hydroxy acids (unidentified).   These products are conjugated to glucosides. *Nocardia* hydrolyze 2,4-D to dichlorophenol (C) which is broken down further by soil microorganisms (D,E, etc.).   *Nocardia* further metab-olize "C" to phenol, monochlorophenols, and chlorocatechol (Menzie, 1966).

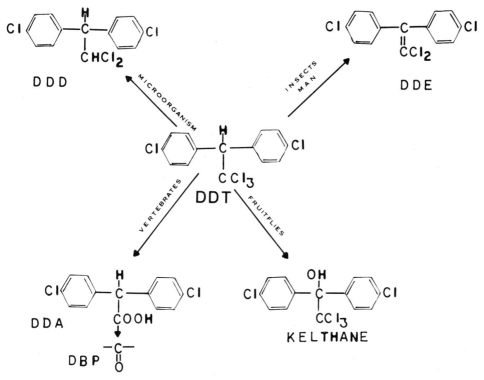

Figure 2. Some of the major detoxication pathways of the insecticide, DDT. Insects, *Daphnia*, a freshwater alga (*Ankistrodesmus amilloides*), and humans dehydrochlorinate it to DDE. Microorganisms dechlorinate it to DDD. Fruit flies and cockroaches oxidize it to kelthane while most vertebrates oxidize it to DDA, dichlorobenzophenone (DBP), dichlorobenzhydrol, etc. Some soil microorganisms (*Hydrogenomonas*) can completely break down DDT to carbon dioxide.

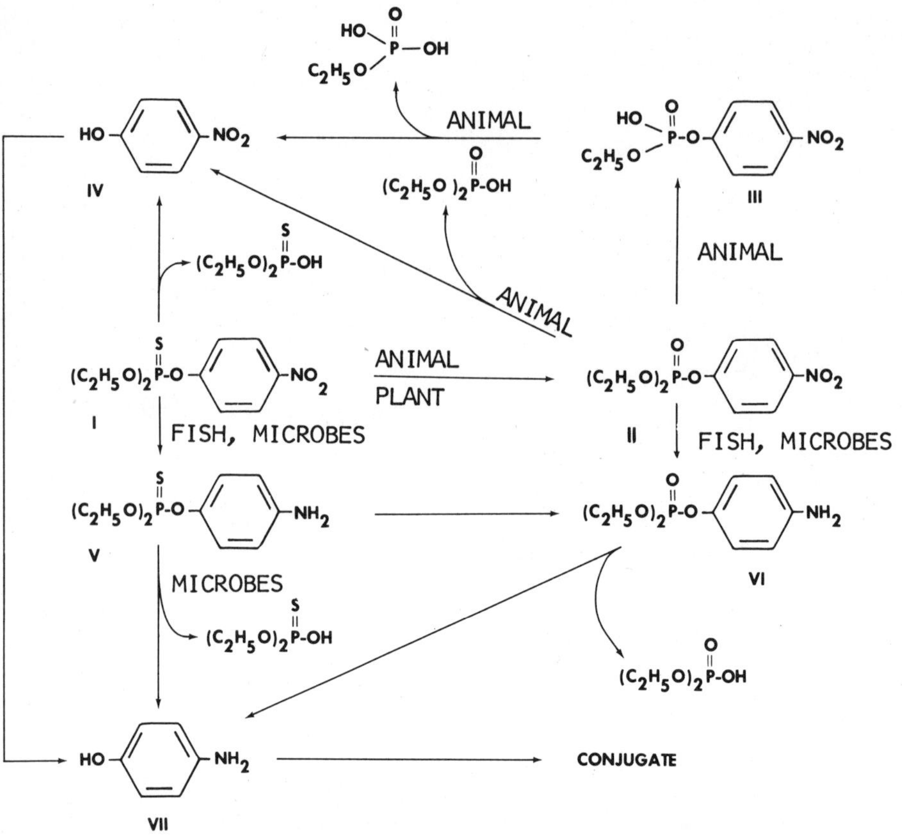

Figure 3.   Metabolism of parathion (I) by biota.   It undergoes rapid oxidative desulfuration in animals and plants to paraoxon (II). Both parathion and paraoxon are reduced in fish and microorganisms to the corresponding amino derivative (V, VI) which may be hydro- lyzed to *p*-aminophenol (VII) and then conjugated.   Animals generally hydrolyze it to *p*-nitrophenol (IV) which may then be reduced. Animals can also de-ethylate paraoxon (III) (Knowles, 1974).

Oxidative dehydrochlorinations.   The photoconversion products of aldrin and dieldrin; namely, photoaldrin and photodieldrin, respectively, are converted by houseflies, mosquito larvae (*Aedes aegypti*), and rats to the dechlorinated photodieldrin ketone (Khan et al., 1970a) (Figure 8 and Table 2).   This ketone is more toxic than either parent compound (Khan et al., 1970b: Khan et al., 1973; Dailey et al., 1972).

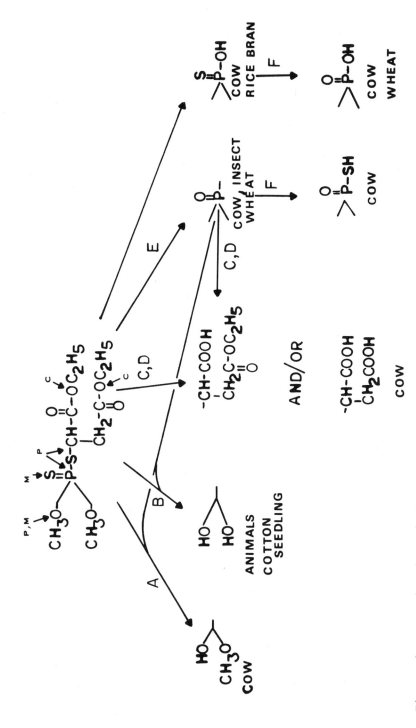

Figure 4. Detoxication of the insecticide malathion by biota. The sites of attack by phosphatase (P), mixed-function oxidase (M), and carboxylesterase (C) can produce desmethyl 'A' or demethyl 'B' products, mono-'C' or diacid 'D' products. Malaoxon 'E' can undergo similar pathways. Phosphatase action can produce phosphates or thiophosphates 'F' which may be des- and demethylated.

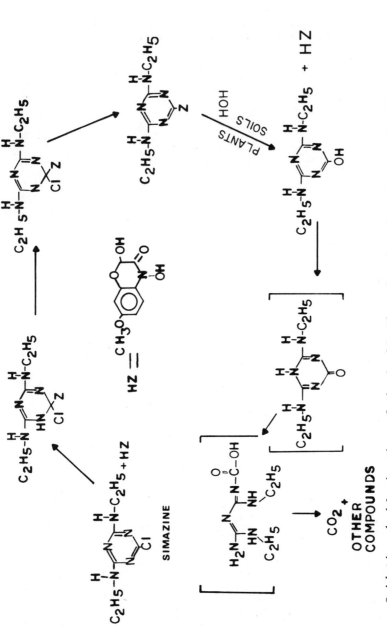

Figure 5. Oxidative dechlorination of the herbicide, simazine, by plants and soil microorganisms.
The fungus, *Aspergillus fumigatus*, can deethylate it first and then oxidatively deaminate it.
Simazine can very slowly undergo ring fission by soil microorganisms. The ring cleavage in higher
plants apparently only occurs after its oxidative dechlorination. In humans, a related derivative
of simazine, formaguanamine (2,4-diamino-s-triazine) is converted to its ammelide (2-amino-4,6-
dihydroxy-s-triazine) (Kearney et al., 1967).

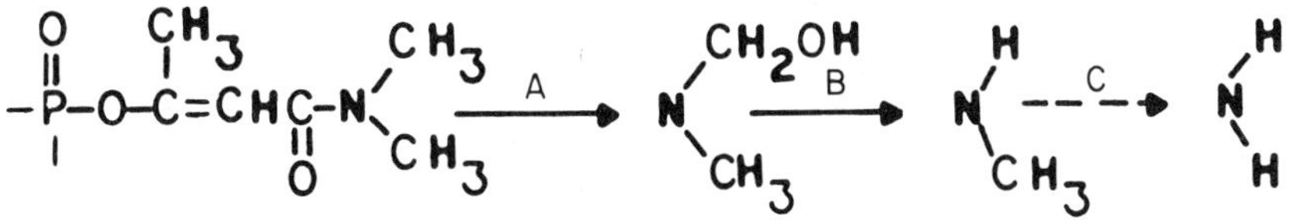

Figure 6. N-dealkylation of the insecticide, Bidrin, which occurs
in plants and animals, via N-methylol (A) formation which can lead
to mono-methyl (B) or demethylated (C) metabolites.

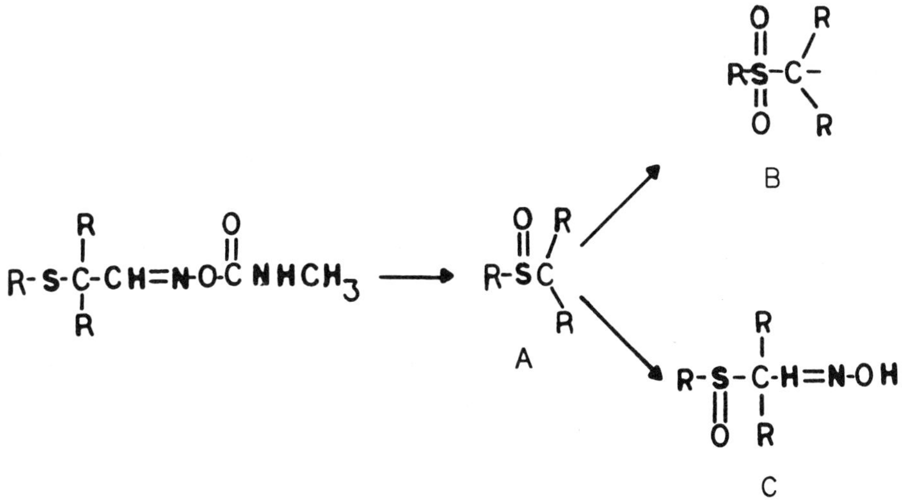

Figure 7. Major pathways of Temik metabolism in plants and
animals. It is converted to its sulfoxide (A) and sulfone (B).
The 'A' may be N-demethylated to sulfoxide oxime (C).

<u>Dechlorination and dehydrochlorination</u>. DDT is anaerobically dechlorinated to DDD by microorganisms in soils and in the gut of animals, and by plants, rats, and mice (Figure 2) (Boush and Batterton, 1972). γ-BHC is dechlorinated by microorganisms to γ-BTC and pentachlorocyclohexane (Figure 9, Table 2) (Ahmed and Casida, 1958; Williams et al., 1972; Tatsukawa et al., 1970).

Chloroaliphatic acids are stable in plants but are dechlorinated by hydrolysis in soil microorganisms (Alexander, 1972). Chlorine substituents on aromatic rings are removed by microorganisms following *ortho*-cleavage of the ring to muconic acids (Figure 1) (Loos et al., 1967; Burger et al., 1962).

The *omega*-substituted phenylalkanoic acids containing a ring chlorine *meta* to the ether oxygen; e.g., 2,4-D and 2,4,5-T, are more resistant to microbial degradation as compared to those with no *meta*-halogen (Ahmed and Casida, 1958; Matsumaka, 1972). However, α substituted phenylalkanoic acids are reasonably persistent regardless of the chlorine on the ring (Alexander, 1972; Coppedge et al., 1967).

Anaerobic dehydrochlorination of DDT to DDE is frequent in terrestrial and aquatic insects (Guenzi and Beard, 1968; Brown, 1971) as well as in the microcrustacean, *Daphnia pulex*, and the alga, *Ankistrodesmus amilloides* (Newdorf and Khan, 1974). It also occurs in mammals including humans (O'Brien, 1967). DDD and *ortho*-chloro-DDT can also be dehydrochlorinated (Guenzi and Beard, 1968). Aerobic dechlorination of the herbicide, simazine, occurs in plants and in soil microorganisms (Figure 5).

Reductions

Reduction of $-NO_2$ groups to $-NH_2$ is common in most organisms but is of special significance in microorganisms and fishes (Table 3). Some common insecticides undergoing this type of reduction are 2,4-dinitro-o-cresol (DNOC), parathion (Figure 3) and EPN (Knowles, 1974; Buhler and Rasmusson, 1968; Henneberg, 1964). The reductive decomposition of the organomercurial fungicides, PMA and MC, by activated sludge has been reported (Figure 10) (Tonomura et al., 1972). *Pseudomonas* strain K-62 reductively decomposes PMA, MC, MMC, and EMP fungicides to metallic mercury (Tonomura et al., 1972). Bacteria (*Arthrobacter simplex* and *Pseudomonas sp.*) reduce $-NO_2$ to $-NH_2$ and replace the latter with -OH in 3,5-DNOC, 2-nitrophenol, and 4-nitrophenol (Faulkner and Woodcock, 1961; Hegenman, 1972; Tonomura et al., 1972).

TABLE 2

Some Examples of Dechlorination Reactions of Pesticides Taking Place in Biota

Detoxication	Pesticide Metabolized		
	Microorganisms	Plants	Animals
Dechlorination:			
Reductive	DDT (Alexander, 1972; Tsukano and Kobayashi, 1972); s-triazine (Harris, 1966); γ-BHC (Allan, 1955)	not common; γ-BHC (Casida and Lykken, 1969)	DDT (Stenersen, 1965); γ-BHC (O'Brien, 1967)
Oxidative	2,4-D (Gutenmann et al., 1964)	2,4-D; 2,4,5-T (Khan and Bederka, 1974; Thomas et al., 1964a)	cyclodienes (Khan et al., 1972b)
Dehydrochlorination:			
Anaerobic	DDT (Alexander, 1972)	not common	DDT (Guenzi and Beard, 1968; Khan and Terriere, 1968)
Oxidative	----	----	photodieldrin (Khan et al., 1970a)

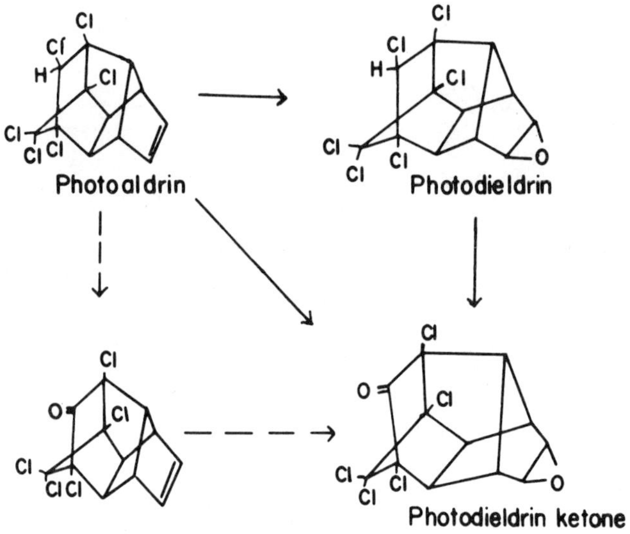

Figure 8. Pathways of oxidative dehydrochlorination of photoaldrin and photodieldrin in insects and possibly in mammals, leading to the formation of the photodieldrin ketone which can also be produced *in vivo* from dieldrin in mammals.

TABLE 3

Some Examples of Reductive Detoxication Reactions of Pesticides

Organisms	Pesticide Metabolized
Microorganism	EPN; denitrothion (Finley and Pillmore, 1963)
Plants	C-6989 (Rogers, 1971); trifluralin (Golab et al., 1967); nitrophen (Kuwatsuka, 1971)
Animals	Parathion (Knowles, 1974); EPN (Buhler and Rasmusson, 1968; Adamson, 1974)

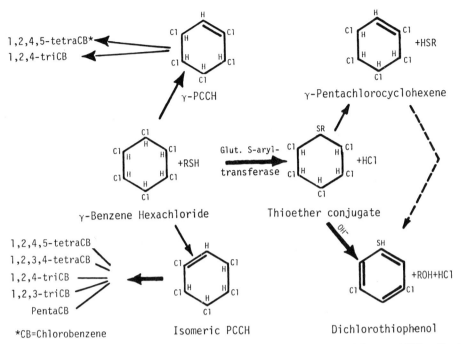

Figure 9. Pathways of metabolism of the insecticide, γ-BHC, (hexachlorocyclohexane) by biota (Perry, 1964).

Hydrolyses

Next to oxidations, hydrolysis constitutes the most important detoxication mechanism, especially since most of the present day pesticides are ester- or amide-type compounds. These ester and amide bonds are easily hydrolyzed by microorganisms, plants, and animals (Table 4). Examples of some of the commonly used pesticides which are hydrolyzed by various microorganisms are seen in Table 5. Plants can hydrolyze 2,4-D to several derivatives, including dichloroanisole (Figure 1). Plants also hydrolyze organophosphate insecticides and fungicides (O'Brien, 1967; Miyamoto, 1972; Uesugi et al., 1972). For example, malathion is converted to dimethyldithiophosphate (Figure 4), dimethylphosphorothiolate, and malathion mono- and diacids (Figure 4). Triazines are degraded by plants (Figure 5) (Hamilton, 1964; Harris et al., 1968). However, plants cannot readily hydrolyze N-alkyl- and N,N'-dialkylcarbamates; e.g., carbaryl and CIPC (Matsumaka, 1972).

Organophosphate and carbamate ester-type insecticides are readily hydrolyzed by insects and vertebrates, and the tolerance of the species depends upon its ability to hydrolyze the particular insecticide.

TABLE 4

Examples of Some of the Pesticides that are Detoxified by Hydrolytic Cleavage of Ester or Amide Bonds

Detoxication	Pesticide Hydrolyzed		
	Microorganisms	Plants	Animals
Ester hydrolysis	dichlorvos, diazinon, parathion, DFP (Boush and Matsumura, 1967); phorate (Ahmed and Casida, 1958); Hinosan (Uesugi et al., 1972)	organophosphate fungicides (Hamilton, 1964) and insecticides (O'Brien, 1967; Miyamoto, 1972); 2,4-D (Miyamoto, 1972; Jooste and Moreland, 1963); carbamate ester insecticides easily hydrolyzed	organophosphate and carbamate insecticides (Casida, 1969; O'Brien, 1967)
Amide hydrolysis	----	propanil (Yih et al., 1968); CMPT (Jooste and Moreland, 1963)	dimethoate (Bull and Adkinson, 1963)

TABLE 5

Hydrolysis of Some Pesticidal Chemicals by Microorganisms
(Matsumura, 1974)

Pesticide	Microorganism
DFP and several organophsophates	*Pseudomonas melapthora* (Boush and Matsumura, 1967)
malathion	*Trichoderma viride* (Matsumura and Boush, 1966)
Thimet	*Pseudomonas fluorescens; Thiobacillus thiooxidans* (Ahmed and Casida, 1958)
Trichlorofon	*Aspergillus niger; Penicillium notatum; Fusarium* (Zayed et al., 1965)
Sumithion	*Bacillus subtilis* (Miyamoto et al., 1966)
Chlorobenzilate	*Rhodotorula gracilis* (Miyazaki et al., 1970)
Zectran	12 spp. (Matsumura, 1974)
Carbaryl	soil microorganisms (Laanio et al., 1972)
Diazinon	soil microorganisms (Laanio et al., 1972)

Amides are readily hydrolyzed by plants and animals. Propanil is readily converted to 2,4-dichloroaniline in rice plants (Matsumaka, 1972). CMPT is hydrolyzed to amino-thiozole in wheat plants (Matsumaka, 1972). The enzymes (amidases) attacking dimethoate are very active in vertebrates and resistant insects but almost inactive in susceptible insects (Smith, 1968).

Ring Cleavage

The benzene ring in certain pesticides; e.g., 2,4-D and DDT, undergoes *ortho*- or *meta*-cleavage in microorganisms (Figure 1) (Alexander, 1972; Loos et al., 1967; Burger et al., 1962). The bacterium *Hydrogenomonas* can utilize DDT as a carbon source (Alexander, 1972; Duxbury et al., 1970). Simazine can also undergo ring fission slowly in soil microorganisms (Figure 5) (Kearney et al., 1967).

Fate of Several Commonly Used Pesticides in Biota

Detoxication of a pesticide by a living organism may involve one or more of the above reactions. Living organisms differ from one another mainly in the rate at which a particular detoxication reaction occurs. However, the qualitative differences in these biotransformations determine the selective toxicity of a pesticide. A particular pesticide; e.g., ester- and amide-type pesticide, can be attacked at several vulnerable sites by more than one detoxifying enzyme. Differences in the activity of these enzymes in various members of the biota can determine its toxicity to target and non-target organisms. For example, malathion can be attacked by various enzymes: carboxyesterases, phosphatases, demethylases, mixed-function oxidases (Figure 4). The mixed-function oxidase (which forms the more toxic malaoxon from malathion) is more active than the carboxy-esterases (which detoxify malathion) in insects, while the reverse is true in mammals; as a result, insects are more sensitive to malathion than are mammals (O'Brien, 1967). Because malathion is attacked by several enzymes, it is rapidly degraded by biota (Figure 4) and, thus, does not persist in the environment for more than a few weeks.

The herbicide, 2,4-D, is detoxified by hydrolysis, hydroxymeth-ylation, dechlorination, decarboxylation, and ring cleavage (Figure 1). Plants convert it to dichloroanisole, decarboxylated metabolites, ester derivatives, and ether-soluble metabolites (Crosby, 1964; Hay and Thimann, 1956; Fang and Butts, 1954). The fungus *Aspergillus niger* converts it to the 2,5-dichloro, 4-hydroxy derivative. The bacteria *Pseudomonas vulgaris* and *Nocardia sp.* convert it to 2,4-dichlorophenol, monochlorophenols, phenol, chlorohydroquinone, chlorocatechol, and several other metabolites (Loos et al., 1967; Taylor and Wain, 1962; Faulkner and Woodcock, 1965). The dichloro-catechol can undergo ring fission to yield α- and β-chloromuconic acids which can be further dechlorinated (Loos et al., 1967; Burger et al., 1962).

The chlorinated hydrocarbon insecticide, DDT, is rapidly de-hydrochlorinated in certain insects (Guenzi and Beard, 1968) but very slowly in the alga, *Ankistrodesmus amilloides*, and the micro-crustacean, *Daphnia pulex* (Neudorf and Khan, 1974), birds, and mammals, including humans (Figure 2) (O'Brien, 1967). Some insects, such as the housefly, *Musca domestica*; the fruit fly *Drosophila melanogaster*; and the cockroach, *Blattella germanica*, oxidize it to Kelthane and other water-soluble metabolites while other organisms, such as the human body louse, *Pediculus humanus humanus*, can oxidize it to DDA (Perry, 1964; Faulkner and Woodcock, 1965; Pinto et al., 1965). A series of metabolites of DDT containing progressively less chlorine can, in turn, give rise to the corresponding partially chlorinated ethylene derivatives. Many of these intermediates can be oxidized by the mixed-function oxidase to secondary metabolites. Hydration of the unsaturated compounds to primary alcohols and their subsequent oxidation leads to DDA formation in rats (O'Brien, 1967). Mammals can, thus, convert DDT to DDE, DDA, dichlorobenzophenone, dichlorobenzhydrol, DDD, and dichlorophenylethane (O'Brien, 1967). Microorganisms in soils and in the gut of animals can reductively dechlorinate it to DDD and DDM (Ahmed and Casida, 1958; Williams et al., 1972; Boush and Batterton, 1972); the former reaction can also be catalyzed by porphyrins (Miskus et al., 1965). Some bacteria can completely break down DDT to carbon dioxide (Alexander 1972; Boush and Batterton, 1972). Fish and plants seem to be unable to metabolize DDT (Neudorf and Khan, 1974; Casida and Lykken, 1969).

Chlorinated polycyclic cyclodiene insecticides are most resistant to biological degradation. The most commonly used of these polychlorinated aliphatic hydrocarbons, such as dieldrin, aldrin, and heptachlor, can persist in soils for several years (Kearney and Kaufman, 1965; Edwards, 1970). Aldrin and dieldrin are converted by sunlight, as well as by microorganisms to their corresponding photoisomers, respectively, photoaldrin and photodieldrin (Rosen and Sutherland, 1967; Matsumura et al., 1970; Khan et al., 1973). These photoproducts are more toxic than their parent compounds to animals at all trophic levels (Khan et al., 1973). Aldrin and dieldrin are also converted to their *trans*-diols in mammals, insects, plants, and fungi (Brooks, 1968; Yu et al., 1971; Korte, 1968; Matthews and Matsumura, 1969; Ludwig and Korte, 1965). In mammals they are converted to photodieldrin ketone (Figure 8) (Klein et al., 1970; Klein et al., 1968). The latter is also produced from photoaldrin and photodieldrin in certain strains of houseflies and mosquitoes (Khan et al., 1969; Harrison et al., 1967; Khan et al., 1970b), rats (Klein et al., 1970), plants and soil (Klein et al., 1973).

Photodieldrin, which is the "terminal residue" of aldrin, dieldrin, and photoaldrin in the environment, is very stable in plants, algae, fish, mammals, and invertebrates. The fresh water algae, *Ankistrodesmus spiralis* and *A. amilloides*, are unable to metabolize photodieldrin in 30 days (Figure 11) (Reddy and Khan, 1974b). However, the fungi, *Aspergillus flavus* and *Penicillium notatum*, convert it to hydrophilic metabolites (Korte, 1970). Male rats convert photodieldrin to photodieldrin ketone and female rats to four hydrophilic metabolites (Klein et al., 1970). The hepatic mixed-function oxidase of rat and mouse converts it *in vitro*, to two and three lipophilic metabolites, respectively; however, the mixed-function oxidase of houseflies forms only one lipophilic metabolite (Figure 11) (Reddy and Khan, 1974a).

The foregoing examples illustrate how some of the most commonly used pesticides can be attacked by various members of the biota and

how their susceptibility or resistance to detoxication enzymes can
determine their fate in the biosphere.

Role of Mixed-Function Oxidase in Detoxication of Cyclodienes

The enzyme systems responsible for the detoxication of pesticides have been well-characterized in insects and mammals. The most
important detoxication system of the biota is the mixed-function
oxidase (MFO) which can catalyze most of the reactions mentioned
under "oxidations, dealkylations, hydrolysis, etc." This system
has been extensively investigated in some mammals in connection with
the metabolism of drugs and insecticides, and in insects in connection with insecticide metabolism. The role of this system in detoxication of pesticides in other organisms is poorly understood.

We have investigated the microsomal metabolism of cyclodiene
insecticides in houseflies, fresh water invertebrates, fresh water
fish, and the pigeon, *Columbia livia*. Two common reactions of these
insecticides; namely, epoxidation and hydroxylation, have been investigated in these comparative studies.

In vivo epoxidation of aldrin (an indicator of MFO activity)
seems to occur widely in algae (Ahmed and Casida, 1958; Williams et
al., 1972), plants (Yu et al., 1971), invertebrates (Table 5) (Khan
et al., 1972a; Khan et al., 1972b), and vertebrates (Korte, 1968).
The rate of *in vivo* epoxidation by fresh water invertebrates is
shown in Table 5. There are differences among animal species in the
initial rate of oxidation of aldrin to dieldrin.

In vitro characterization of this system in fresh water invertebrates has shown that the activity is generally localized in the
hepatopancreas or gut of the crayfish (*Cambarus*), snail (*Lymnaea*),
and the mussel (*Anodonta*) (Khan et al., 1972a; Khan et al., 1972b).
The epoxidation and hydroxylation of cyclodienes by the MFO in fresh
water invertebrates and fish, pigeon, and mouse is very sensitive
to changes in temperature and pH. The optimum activity of MFO occurs
at pH 7.3 to 8.4 and at 26°C in invertebrates (Table 6). The MFO
of the cold water fish, bass and bluegill, also show maximum activity
at 26°C but the pH optima are at 7.4 and 8.0-8.2, respectively
(Stanton and Khan, 1973). The MFO of the tropical fish, Kissing
Gourami, shows maximal activity at 37°C and at pH 7.6 (Garreto and
Khan, 1974). Pigeon MFO is most active around 45°C and pH 6.8
(Runnels and Khan, 1973), while the mouse enzyme is active at 37°C
and pH 7.4 (Stanton and Khan, 1973).

Inter- and intra-specific differences in the rate of *in vitro*
epoxidation and hydroxylation of chlordene have also been observed
(Stanton and Khan, 1973; Garretto and Khan, 1974; Runnels and Khan,

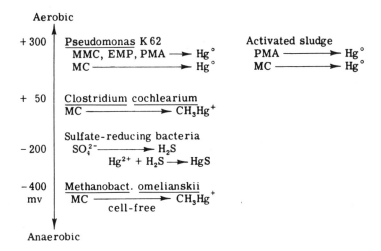

Figure 10.   Pathways of metabolism of the organomercurial pesticides, PMA and MC, by microorganisms (Tonomura et al., 1972).

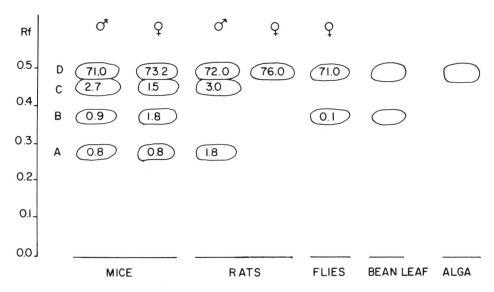

Figure 11.   Thin-layer chromatographic presentation of metabolic products formed *in vitro* by the mixed function oxidase of various animals and in bean leaves in the presence of sunlight.   The numbers in each spot indicate the percentage of the total radioactivity (Reddy and Khan, 1974a; Reddy and Khan, 1974b).

TABLE 6

Pick-up of Aldrin and Its Subsequent Epoxidation by Freshwater Invertebrates
During Four-Hour Exposure to .1 ppm Aldrin (Khan et al., 1972a)

| Animal | Insecticide Absorbed: ng/Animal | | | Epoxidation of Aldrin |
	Aldrin	Dieldrin	Total	% Dieldrin/Animal
1. Hydra littoralis (Coelenterate)	45.02	1.20	46.25	2.59
2. Dugesia (Planarian)	23.01	0.24	23.25	1.03
3. Leech (Annelid)	34.14	0.54	35.65	1.51
4. Asellus (Crustacea, isopod)	25.02	0.59	25.61	2.30
5. Gammarus (Crustacea, amphipod)	19.98	1.52	21.50	7.07
6. Daphnia pulex (Crustacea, Cladocera)	27.16	1.64	28.80	5.69
7. Cyclops (Crustacea, copepoda)	26.23	1.13	27.36	4.13

8. *Cambarus* (Crustacea, Decapoda)	86.43	8.01	94.44	8.48
9. *Aeschna* (Insecta, Odonata)	48.59	16.11	64.70	24.90
10. *Aedes aegypti* (Insecta, Diptera)	95.06	69.82	164.88	42.35
11. *Anodonta* (Mollusk, pelecypoda)	553.89	130.81	684.70	19.10
12. *Lymnaea palustis* (Mollusk, Gastropoda)	69.15	13.95	83.10	16.79

1973). Under optimum conditions, the MFO of mouse and pigeon liver is more active than that of fish in oxidizing chlordene or aldrin (Table 7).

In vitro metabolism of cyclodienes by housefly and rat MFO systems has been studied by Brooks and coworkers (Brooks, 1969. Species differences in epoxidation, hydroxylation, and epoxide hydration are quite obvious (Khan, 1973). Sesamex-sensitive *in vitro* metabolism of dieldrin by rat liver MFO amounts to only 9%, and the major metabolite is monohydroxy-photodieldrin which is subsequently conjugated with glucuronic acid (Matthews et al., 1971). Photodieldrin is metabolized by the mammalian and insect MFO to ether-soluble metabolites. Three such metabolites were found in male mouse, two in male rat, and only one in the housefly (Figure 11) (Reddy and Khan, 1974a). The MFO of certain strains of houseflies also converts photodieldrin to hydrophilic metabolites (Khan et al., 1970a).

Conjugation of Pesticide Metabolites

The polar metabolite(s) resulting from the initial detoxication of a pesticide may or may not be conjugated prior to excretion. Microorganisms do not perform conjugation reactions because excretion takes place through the cell surface.

In contrast to microorganisms and animals, plants cannot excrete pesticides and their metabolites. Instead, they conjugate these foreign chemicals with endogenous compounds (Table 8) and deposit them in metabolically inactive sites in the cell (e.g., vacuoles). Polymerization of the conjugate may also occur. Conjugation may facilitate the transport of these conjugates between tissues in higher plants as well as in animals. Animal groups differ in their choice of conjugation reactions but there are more similarities than differences among members of related groups.

Various synthetic reactions of pesticide detoxication products are seen in Table 8 and Figure 12. Although the conjugation systems are more rapid than the initial detoxication reactions (Klein et al., 1968; Williams, 1959) in plants and animals, the latter phase is more important in pesticide toxicity even in organisms where these detoxication reactions occur at very slow rates. Thus, the knowledge of the detoxication of pesticides can only help to evaluate the health and ecological effects of present-day synthetic pesticides.

SUMMARY

The current status of knowledge of the detoxication of pesticidal chemicals by biota has been examined. Pathways of detoxica-

TABLE 7

Temperature and pH Optima for Maximal Rate of Epoxidation and Hydroxylation of Cyclodienes by the MFO of Various Animals

Animal	Reaction[a]	Optima pH	Optima Temp., °C	Reference
Bass fry	epoxidation and hydroxylation	8-8.2	26	Stanton and Khan, 1973
Bluegill young adult	epoxidation and hydroxylation	7.4	26	Garretto and Khan, 1974
Kissing Gourami	epoxidation and hydroxylation	7.6	37	Garretto and Khan, 1974
Trout	epoxidation and hydroxylation	7.4	--	Garretto and Khan, 1974
Crayfish: green gland	epoxidation	7.8	26	Khan et al., 1972a / Khan et al., 1972b
hepatopancreas	epoxidation	7.3	26	Khan et al., 1972a / Khan et al., 1972b
Snail	epoxidation		29	Khan et al., 1972a / Khan et al., 1972b
Clam	epoxidation		29	Khan et al., 1972a / Khan et al., 1972b
Pigeon (male)	epoxidation and hydroxylation	6.8-7.0	45	Runnels and Khan, 1973
Mouse (male)	epoxidation and hydroxylation	7.4	37	Stanton and Khan, 1973
Houseflies (female)	epoxidation	7.8	34	Khan et al., 1970

[a] Epoxidation of aldrin; and epoxidation and hydroxylation of chlordene.

TABLE 8

Rates of Epoxidation and Hydroxylation of Cyclodienes by
MFO of Various Animals Under Optimum Conditions

p-mole Product/Min/Mg Protein

Animal	Aldrin → Dieldrin	Photoaldrin → Photodieldrin	Isodrin → Endrin	Chlordene Oxidation[a]			
				CE	COH	CEOH	Total
Bass fry	1.19	--	--	--	--	--	--
Bluegill fry	1.45	--	0.81	0.13	0.99	--	1.12
Bluegill young adult	--	--	--	0.14	1.01	--	1.15
Kissing Gourami	1.86	2.50	0.73	0.14	0.34	0.09	0.57
Mouse (male)	3.35	2.15	1.17	0.23	1.13	--	1.36
Pigeon (male)	1.43	--	--	0.26	0.44	0.10	0.80

[a] Chlordene oxidation products are: 2,3-epoxychlordene (CE); 1-hydroxychlordene (COH), and 1-hydroxy-2,3-epoxychlordene (CEOH); sum total of these three metabolites shown in last column.

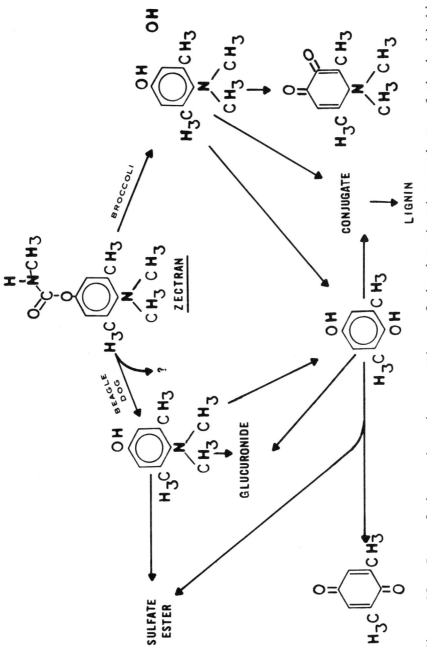

Figure 12. Some of the conjugation reactions of the detoxication products of the herbicide, Zectran, in plants and animals (Menzie, 1966).

TABLE 9

Conjugating Compound	Reactive Group(s) of Pesticide(s)	Pesticide Conjugated	
		Plants	Animals
Glucose	hydroxyl, carboxyl, thiol, amide, amine	dimethyldithiocarbamates (Casida and Lykken, 1969; Sijperstein and Kaslander, 1964; Stromme, 1965); N-methylcarbamates (Kuhr and Casida, 1967); arylacetic and phenoxyacetic acids; herbicides with amino groups (amiben), substituted anilines, etc. (Still, 1968; Lamoureux et al., 1971)	insects: N-methyl and N,N'-dimethylcarbamates, dimethyldithiocarbamates (Sijperstein and Kaslander, 1964; Stromme, 1965)
Glucuronic acid	hydroxyl, carboxyl, thiol, amide, amine	-----	vertebrates (Adamson, 1974)
Glutathione	halogens	atrazine (Shimabukuro et al., 1970); propaclor (Lamoureux et al., 1971)	dichloronitrobenzene (Grover and Sims, 1964); PCCH (Clark et al., 1966)
amino acids	carboxyl	2,4-D and amitrole (Feung et al., 1971); dimethyl-dithiocarbamates (Casida and Lykken, 1969)	DDA in rats (Matsumura, 1974)
Sulfate	phenols	absent	form N-sulfates and phenylsulfamic acids (Smith, 1968)

| Acetylation[a] and Methylation[a] | amines, phenols, catechols | phenols or catechols (Smith, 1968) | acetylation of amines, formatenate; methylation of substituted catechols and pyridine derivatives (Smith, 1968) |

[a] Very important in microorganisms especially with organomercurials and arylamines.

tion include initial metabolic alterations in target organisms, such as oxidations (oxidative desulfuration, dealkylation, decarbamylation, N-dealkylation, aliphatic and aromatic hydroxylation, sulfoxide and sulfone formation, etc.) and reductions (dehydrochlorination, dechlorinated, nitro-reduction, etc.). Further metabolism and conjugation of these initially altered molecules leading to their excretion has been studied in target organisms and those at the top of the food chain including humans. Differences in the pathways of metabolism of some extensively used cyclodiene pesticides in nontarget organisms at various trophic levels is considered. This review of *in vivo* and *in vitro* metabolism emphasizes the role of detoxication in predicting the fate and effects of pesticides and their metabolites on biota may be used in the design, development, and usage of successful future pesticides or in the modification of those used presently.

Acknowledgment

    Support by Grant ES 00808 (National Institute of Environmental Health Sciences).

REFERENCES

Adamson, R.H. 1974. Survival in Toxic Environments. M.A.Q. Khan
    and J.P. Bederka, Jr., editors. Academic Press, New York, 550
    pp.

Ahmed, M.K., and J.E. Casida. 1958. Metabolism of some organophos-
    phorus insecticides by microorganisms. J. Econ. Entomol. 51:
    59.

Alexander, M. 1972. Environmental Toxicology of Pesticides.
    F. Matsumura, G.M. Boush, and T. Misato, editors. Academic
    Press, New York, 637 pp.

Allan, J. 1955. Loss of biological efficiency of cattle-dipping
    wash containing benzene hexachloride. Nature 175:1131.

American Chemical Society. 1973. Significance of Pesticide Metabo-
    lites. A Symposium. 166th National Meeting, Chicago, Illinois.

Andrews, N.R., H.W. Dorough, and D.A. Lindquist. 1967. Degradation
    and elimination of Temik in rats. J. Econ. Entomol. 60:979.

Beneset, H., and F. Matsumura. Unpublished data.

Blinn, R.C. 1968. Abate insecticide: the fate of O,O,O',O'-
    tetramethyl O,O'-thiodi-p-phenylene phosphorothioate on bean
    leaves. J. Agr. Food Chem. 16:441.

Bollag, J.M., and S.Y. Liu. 1971. Degradation of Sevin by soil
    microorganisms. Soil Biol. Biochem. 3:337.

Boush, G.M., and J.C. Batterton. 1972. Environmental Toxicology
    of Pesticides. F. Matsumura, G.M. Boush, and T. Misato,
    editors. Academic Press, New York, 637 pp.

Boush, G.M., and F. Matsumura. 1967. Insecticidal degradation by
    pseudomonas melophthora, the bacterial symbiote of the apple
    maggot. J. Econ. Entomol. 60:918.

Bowman, M.C., M. Beroza, and J.A. Harding. 1969. Determination
    of phorate and five of its metabolites in corn. J. Agr. Food
    Chem. 17:138.

Brooks, G.T. 1968. Symposium on the Science and Technology of
    Residual Insecticides in Food Production With Special Reference
    to Aldrin and Dieldrin. Shell Oil Company, USA, 244 pp.

Brooks, G.T. 1969. The metabolism of diene-organochlorine (cyclo-diene) insecticides. Residue Reviews 27:81.

Brooks, G.T., A. Harrison, and S.E. Lewis. 1970. Cyclodiene epoxide ring hydration by microsomes from mammalian liver and houseflies. Biochem. Pharmacol. 19:255.

Brown, A.W.A. 1971. Pesticides in the Environment. L. White-Stevens, editor. Marcel Dekker, p. 457.

Buhler, D.R., and M.E. Rasmusson. 1968. Reduction of p-nitroben-zoic acid by fishes. Arch. Biochem. Biophys. 124:582.

Bull, D.L., and P.L. Adkinson. 1963. Absorption and metabolism of C^{14}-labeled DDT by DDT-susceptible and DDT-resistant pink bollworm adults. J. Econ. Entomol. 56:641.

Bull, D.L., and D.A. Lindquist. 1964. Metabolism of 3-hydroxy-N,N-dimethyl-crotonamide dimethyl phosphate by cotton plants, insects, and rats. J. Agr. Food Chem. 12:310.

Bull, D.L. 1965. Metabolism of di-syston by insects, isolated cotton leaves, and rats. J. Econ. Entomol. 58:249.

Burger, K., I.C. MacRae, and M. Alexander. 1962. Decomposition of phenoxyalkyl carboxylic acids. Soil. Sci. Soc. Amer. Proc. 26:243.

Casida, J.E. 1969. Microsomes and Drug Oxidations. J.R. Gillette, A.H. Conney, G.J. Cosmides, R.W. Estabrook, J.R. Fonts, and G.J. Mannering, editors. Academic Press, 547 pp.

Casida, J.E., and L. Lykken. 1969. Metabolism of organic pesticide chemicals in higher plants. Ann. Rev. Plant Physiol. 20:607.

Chakraborty, J., and J.N. Smith. 1967. Enzymic oxidations of some alkylbenzenes in insects and vertebrates. Biochem. J. 102:498.

Chapman, P.J. 1972. Degradation of Synthetic Organic Molecules in the Biosphere. Nation. Acad. Sci., 347 pp.

Clark, A.G., M. Hitchcock, and J.N. Smith. 1966. Metabolism of gammexane in flies, ticks, and locusts. Nature 209:103.

Clayson, D.B., and M.J. Ashton. 1963. The metabolism of 1-naph-thylamine and its bearing on the mode of carcinogenesis of the aromatic amines. Acta Unio Intern. Contra. Cancrum 19:539.

Coppedge, J.R., D.A. Lindquist, D.L. Bull, and H.W. Dorough. 1967. Fate of 2-methyl-2-(methylthio) propionaldehyde O-(methylcar-bamoyl) oxime (Temik) in cotton plants and soil. J. Agr. Food Chem. 15:902.

Crosby, D.G. 1964. Metabolites of 2,4-dichlorophenoxyacetic acid (2,4-D) in bean plants. J. Agr. Food Chem. 12:3.

Dailey, R.E., A.K. Klein, E. Brouwer, J.D. Link, and R.C. Braunberg. 1972. Effect of testosterone on metabolism of ^{14}C-photodieldrin in normal, castrated, and oophorectomized rats. J. Agr. Food Chem. 20:371.

Dorough, H.W. and J.E. Casida. 1964. Nature of certain carbamate metabolites of the insecticide Sevin. J. Agr. Food Chem. 12: 294.

Duxbury, J.M., J.M. Tiedje, M. Alexander, and J.E. Dawson. 1970. 2,4-D metabolism: Enzymatic conversion of chloromaleylactic acid to succinic acid. J. Agr. Food Chem. 18:199.

Edwards, C.A. 1970. Persistence Pesticides in the Environment. The Chem. Rubber Co. Press, 78 pp.

Eldefrawi, M.E., and W.M. Hoskins. 1961. Relation of the rate of penetration and metabolism to the toxicity of Sevin to three insect species. J. Econ. Entomol. 54:401.

Fang, S.C., and J.S. Butts. 1954. Studies in plant metabolism. III. Absorption, translocation, and metabolism of radioactive 2,4-D in corn and wheat plants. Plant Physiol. 29:56.

Faulkner, J.K., and D. Woodcock. 1961. Agri. Biol. Chem. J. 70: 373.

Faulkner, J.K., and D. Woodcock, 1965. Fungal detoxication. Part VII. Metabolism of 2,4-dichlorophenoxyacetic and 4-chloro-2-methylphenoxy-acetic acids by Aspergillus niger. J. Chem. Soc. 1187.

Fawcett, C.H., J.M.A. Ingram, and R.L. Wain. 1954. The β-oxidation of ω-phenoxyalkylcarboxylic acids in the flax plant in relation to their plant growth-regulating activity. Proc. Royal Soc. London B142:60.

Feung, C.S., R.H. Hamilton, and F.H. Witham. 1971. Metabolism of 2,4-dichlorophenoxyacetic acid by soybean cotyledon callus tissue cultures. J. Agr. Food Chem. 19:475.

Finley, R.B., and R.E. Pillmore. 1963. Conversion of DDT to DDD in animal tissue. Am. Inst. Biol. Sci. Bull. 13:41.

Garretto, M., and M.A.Q. Khan. 1974. Mixed-function oxidase in fresh water fish: Bluegill and Kissing Guorami. Paper presented at the 7th annual meeting of Illinois State Acad. Sci. submitted to Comp. Biochem. Physiol.

Gillette, J.R., A.H. Conney, G.J. Cosmides, R.W. Estabrook, J.R. Fout, and G.J. Mannering. 1969. Microsomes and Drug Oxidations. Academic Press, New York, 547 pp.

Golab, T., R.J. Herberg, S.J. Parka, and J.B. Tepe. 1967. Metabolism of carbon-14 trifluralin in carrots. J. Agr. Food Chem. 15:638.

Grover, P.L., and P. Sims. 1965. The metabolism of γ-2,3,4,5,6-pentachlorocyclohex-1-ene and γ-hexachlorocyclohexane in rats. Biochem. J. 96:521.

Guenzi, W.D., and W.E. Beard. 1968. Anaerobic conversion of DDT to DDE and aerobic stability of DDT in soil. Soil Sci. Soc. Amer. Proc. 32:522.

Gutenmann, W.H., M.A. Loos, M. Alexander, and D.J. Lisk. 1964. Beta oxidation of phenoxyalkanoic acids in soil. Soil Sci. Soc. Amer. Proc. 28:205.

Hamilton, R.H. 1964. Tolerance of several grass species to 2-chloro-s-triazine herbicides in relation to degradation and content of benzoxazinone derivatives. J. Agr. Food Chem. 12: 14.

Harris, C.R. 1966. Influence of soil type on the activity of insecticides in soil. J. Econ. Entomol. 59:1221.

Harris, C.I., D.D. Kaufman, T.J. Sheets, R.G. Nash, and P.C. Kearney. 1968. Behavior and fate of s-triazines in soils. In Advances in Pest Control. R.L. Metcalf, editor, 8:1.

Harrison, R.B., D.C. Holmes, J. Roburn, and Z.O'G. Tatton. 1967. The fate of some organochlorine pesticides on leaves. J. Sci. Fd. Agric. 18:10.

Hay, J.R., and K.V. Thimann. 1956. The fate of 2,4-dichlorophenoxyacetic acid in bean seedlings. I. Recovery of 2,4-dichlorophenoxyacetic acid and its breakdown in the plant. Plant Physiol. 31:382.

Hegenman, G.W. 1972. Degradation of Synthetic Organic Molecules in the Biosphere. Nation. Acad. Sci. 347 pp.

Henneberg, M. 1964. Dinitroisopiopylphenol (DNPP) and dinitrobutyl-phenol (DNBP) metabolites in rats. Acta Poloniae Pharmacentics 21:222.

Hodgson, E. 1969. Enzymatic Oxidation of Toxicants. North Carolina State University at Raleigh, 228 pp.

Institute fur Okologische Chemie. 1970. Metabolism von Pestiziden und Ihr Vehallen unter Umweltbedingungen. Internationale Symposium, Bonn, Germany.

Jooste, J., and D.E. Moreland. 1963. Preliminary characterization of some plant carboxylic ester hydrolases. Phytochemistry 2: 263.

Kearney, P.C., and D.D. Kaufman. 1965. Enzyme from soil bacterium hydrolyzes phenylcarbamate herbicides. Science 147:740.

Kearney, P.C., D.D. Kaufman, and M. Alexander. 1967. Soil Bio-chemistry. A. McLaren and G.H. Peterson, editors. Marcel Dekker.

Khan, M.A.Q., and J.P. Bederka, Jr. 1974. Survival in Toxic Environments. Academic Press, 550 pp.

Khan, M.A.Q., J.L. Chang, D.J. Sutherland, J.D. Rosen, and A. Kamal. 1970a. House fly microsomal oxidation of some foreign compounds. J. Econ. Entomol. 63:1807.

Khan, M.A.Q., W.F. Coello, A.A. Khan, and H. Pinto. 1972a. Some characteristics of the microsomal mixed-function oxidase in the freshwater crayfish, *Cambarus*. Life Sci. 11:405.

Khan, M.A.Q., and W.O. Hauge. 1970. Toxicology, Biodegradation, and Efficacy of Pesticides. Swets and Zeitlinger, 434 pp.

Khan, M.A.Q., A. Kamal, R.J. Wolin, and J. Runnels. 1972b. *In vivo* and *in vitro* epoxidation of aldrin by aquatic food chain organisms. Bull. Environ. Contam. Toxicol. 8:219.

Khan, M.A.Q., J.D. Rosen, and D.J. Sutherland. 1969. Insect metabolism of photoaldrin and photodieldrin. Science 164: 318.

Khan, M.A.Q., R.H. Stanton, and G. Reddy. 1974. Survival in Toxic Environments. M.A.Q. Khan and J.P. Bederka, Jr., edi-tors. Academic Press, 550 pp.

Khan, M.A.Q., R.H. Stanton, D.J. Sutherland, J.D. Rosen, and N. Maitra. 1973. Toxicity metabolism relationship of the photoisomers of cyclodiene insecticides. Arch. Environ. Contam. Toxicol. 1:159.

Khan, M.A.Q., D.J. Sutherland, J.D. Rosen, and W. Carey. 1970b. Effect of sesamex on the toxicity and metabolism of cyclodienes and their photoisomers in the house fly. J. Econ. Entomol. 63:470.

Khan, M.A.Q., and L.C. Terriere. 1968. DDT-dehydrochlorinase activity in house fly strains resistant to various groups of insecticides. J. Econ. Entomol. 61:732.

Klein, A.K., R.E. Dailey, M.S. Walton, V. Beck, J.D. Link. 1970. Metabolites isolated from urine of rats fed ^{14}C-photodieldrin. J. Agr. Food Chem. 18:705.

Klein, A.K., J.D. Link, N.F. Ives. 1968. Isolation and purification of metabolites found in the urine of male rats fed aldrin and dieldrin. J. Assoc. Off. Anal. Chem. 51:895.

Klein, W., J. Kohli, I. Weisgerber, and F. Korte. 1973. Fate of aldrin-^{14}C in potatoes and soil under outdoor conditions. J. Agr. Food Chem. 21:152.

Knaak, J.B., M.J. Tallant, W.J. Bartley, and L.J. Sullivan. 1965. The metabolism of carbaryl in the rat, guinea pig, and man. J. Agr. Food Chem. 13:537.

Knowles, C.O. 1974. Survival in Toxic Environments. M.A.Q. Khan and J.P. Bederka, Jr., editors. Academic Press, 550 pp.

Korte, F. 1967. Metabolism of ^{14}C-labeled insecticides in microorganisms, insects, and mammals. Botgu-Kagaku 32:46.

Korte, F. 1968. Symposium on the Science and Technology of Residual Insecticides in Food Production With Special Reference to Aldrin and Dieldrin. Shell Oil Company, USA, 244 pp.

Korte, F. 1970. IUPAC Commission on Terminal Residues. J. Ass. Off. Analy. Chem. 53:987.

Kuhr, R.J. 1970. Metabolism of carbamate insecticide chemicals in plants and insects. J. Agr. Food Chem. 18:1023.

Kuhr, R.J., and J.E. Casida. 1967. Persistent glycosides of metabolites of methylcarbamate insecticide chemicals formed by hydroxylation in bean plants. J. Agr. Food Chem. 15:814.

Kuwatsuka, S. 1971. Environmental Toxicology of Pesticides. F. Matsumura, G.M. Boush, and T. Misato, editors. Academic Press, 637 pp.

Laanio, T.L., G. Dupuis, and H.O. Esser. 1972. Fate of ^{14}C-labeled diazinon in rice, paddy soil, and pea plants. J. Agr. Food Chem. 20:1213.

Lamoureux, G.L., L.E. Stafford, and F.S. Tanaka. 1971. Metabolism of 2-chloro-N-isopropylacetanilide (propachlor) in the leaves of corn, sorghum, sugarcane, and barley. J. Agr. Food Chem. 19:346.

Leeling, N.C., and J.E. Casida. 1966. Metabolites of carbaryl (1-naphthyl methylcarbamate) in mammals in enzymatic systems for their formation. J. Agr. Food Chem. 14:281.

Lemin, A.J. 1966. Absorption, translocation, and metabolism of diphenamid-1-C^{14} by tomato seedlings. J. Agr. Food Chem. 14: 109.

Liu, S.Y., and J.M. Bollag. 1971. Metabolism of carbaryl by a soil fungus. J. Agr. Food Chem. 19:487.

Loos, M.A., R.N. Roberts, and M. Alexander. 1967. Phenols as intermediates in the decomposition of phenoxyacetates by an arthrobacter species. Can. J. Microbiol. 13:679.

Ludwig, G., and F. Korte. 1965. Metabolism of insecticides. X. Detection of dieldrin metabolite by GLC analysis. Life Sci. 4:2027.

Matsumura, F. 1974. Survival in Toxic Environments. M.A.Q. Khan and J.O. Bederka, Jr., editors. Academic Press, 550 pp.

Matsumura, F., and G.M. Boush. 1966. Malathion degradation by trichoderma viride and a pseudomonas species. Science 153:1278.

Matsumura, F., G.M. Boush, and T. Misato. 1972. Environmental Toxicology of Pesticides. Academic Press, 637 pp.

Matsumura, F., K.S. Patil, and G.M. Boush. 1970. Formation of "photodieldrin" by microorganisms. Science 170:1206.

Matsunaka, S. 1972. Environmental Toxicology of Pesticides. F. Matsumura, G.M. Boush, and T. Misato, editors. Academic Press, 637 pp.

Matsunaka, S., and H. Nakamura. 1972. Mode of action and selectivity mechanism of a herbicide, 5-chloro-4-methyl-2-propionamide-1,3-thiazole. Weed Res. (Tokyo) 13:29.

Matthews, H.B., and F. Matsumura. 1969. Metabolic fate of dieldrin in the rat. J. Agr. Food Chem. 17:845.

Matthews, H.B., J.D. McKinney, and G.W. Lucier. 1971. Dieldrin metabolism, excretion, and storage in male and female rats. J. Agr. Food Chem. 19:1244.

Menzer, R.E., and J.E. Casida. 1965. Nature of toxic metabolites formed in mammals, insects, and plants from 3-(dimethoxphos-phinyloxy)-N,N-dimethyl-cis-crotonamide and its N-methyl analog. J. Agr. Food Chem. 13:102.

Menzie, C.M. 1966. Metabolism of pesticides. U.S. Department of Interior, Fish and Wildlife Service, Special Scientific Rep. No. 96 (Wildlife). 274 pp.

Miskus, R.P., D.P. Blair, and J.E. Casida. 1965. Conversion of DDT to DDD by bovine rumen fluid, lake water, and reduced porphyrins. J. Agr. Food Chem. 13:481.

Miyamoto, J. 1972. Environmental Toxicology of Pesticides. F. Matsumura, G.M. Boush, and T. Misato, editors. Academic Press, 637 pp.

Miyamoto, J., A. Kitagawa, and Y. Sato. 1966. Metabolism of organo-phosphorus insecticides by *Bacillus subtilis* with special em-phasis on Sumithion. Jap. J. Expt. Med. 36:211.

Miyazaki, S., G.M. Boush, and F. Matsumura. 1970. Microbial de-gradation of chlorobenzilate (ethyl 4,4'-dichlorobenzilate) and chloropropylate (isopropyl 4,4'-dichlorobenzilate). J. Agr. Food Chem. 18:87.

Nashed, R.B., S.E. Katz, and R.D. Illnicki. 1970. The metabolism of ^{14}C-chlorbromuron in corn and cucumber. Weed Sci. 18:122.

National Academy of Sciences. 1972. Degradation of synthetic organic molecules in the biosphere, Proceedings of a conference in San Francisco, June, 1971, 350 pp.

Neudorf, S., and M.A.Q. Khan. 1974. Paper presented at the 7th annual meeting of the Illinois State Academy of Sciences, sub-mitted to Bull. Environ. Contam. Toxicol.

O'Brien, R.D.   1962.   Metabolic Factors Controlling Duration of Drug
    Action.   B.B. Brodie and E.G. Erdos, editors.   McMillan
    (Pergammon) Press, p. 111.

O'Brien, R.D.   1967.   Insecticides: Action and Metabolism.   Aca-
    demic Press, 378 pp.

O'Brien, R.D. and I. Yamamoto.   1970.   Biochemical Toxicology of
    Insecticides.   Academic Press, 218 pp.

Perry, A.S.   1964.   The physiology of insecticide resistance by
    insects.   Physiology of Insects.   M. Rockstein, editor.
    Academic Press, Vol. 3:286.

Pinto, J.D., M.N. Camien, and M.S. Dunn.   1965.   Metabolic fate of
    p,p'-DDT[1,1,1-trichloro-2,2-bis(p-chlorophenyl)ethane] in
    rats.   J. Biol. Chem. 240:2148.

Reddy, G., and M.A.Q. Khan.   1974a, in press.   *In vitro* metabolism
    of ^{14}C-photodieldrin by microsomal mixed function oxidase of
    mice, rat, and houseflies.   J. Agr. Food Chem.

Reddy, G., and M.A.Q. Khan.   1974b, in press.   Persistance of ^{14}C-
    photodieldrin under various environmental conditions.   Bull.
    Environ. Contam. Toxicol.

Rogers, R.L.   1971.   Absorption, translocation, and metabolism of
    p-nitrophenyl-α,α,α-trifluoro-2-nitro-p-tolyl ether by soy-
    beans.   J. Agr. Food Chem. 19:32.

Rosen, J.D., and D.J. Sutherland.   1967.   The nature and toxicity
    of the photoconversion products of aldrin.   Bull. Environ.
    Contam. Toxicol. 2:1.

Runnels, J.M., and M.A.Q. Khan.   1973.   Hepatic mixed-function
    oxidase activity towards cyclodiene insecticides in the domes-
    tic pigeon.   Amer. Zool. 13:1308.

Shimabukuro, R.H., H.R. Swanson, and W.C. Walsh.   1970.   Atrazine
    detoxication mechanism in corn.   Plant Physiol. 46:103.

Sijperstein, A.K., and J. Kaslander.   1964.   Metabolism of fungi-
    cides by plants and microorganisms.   Outlook Agr. 4:119.

Smith, J.N.   1968.   The comparative metabolism of xenobiotics.   In
    Advances in Comparative Biochemistry and Physiology.   O.
    Lowenstein, editor.   3:173.

Stanton, R.H., and M.A.Q. Khan. 1973. Mixed-function oxidase activity toward cyclodiene insecticides in bass and bluegill sunfish. Pest. Biochem. Physiol. 3:351.

Stenersen, J.H.V. 1965. DDT-metabolism in resistant and susceptible stableflies and in bacteria. Nature 207:660.

Still, G.G. 1968. Metabolism of 3,4-dichloropropionanilide in plants: The metabolic fate of the 3,4-dichloroaniline moiety. Science 159:992.

Stromme, J.H. 1965. Metabolism of disulfiram and diethyldithiocarbamate in rats with demonstration of an *in vivo* ethanol-induced inhibition of the glucuronic acid conjugation of the thiol. Biochem. Pharmacol. 14:393.

Tatsukawa, R., T. Wakimoto, and T. Ogawa. 1970. J. Food Hyg. Soc. (Japan) 11:1.

Taylor, H.F., and R.L. Wain. 1962. Side-chain degradation of certain ω-phenoxyalkane carboxylic acids by *Nocardia coeliaca* and other microorganisms isolated from soil. Proceed. Royal Soc. London (Series B) 156:172.

Thomas, E.W., B.C. Loughman, and P.G. Powell. 1964a. Metabolic fate of some chlorinated phenoxyacetic acids in the stem tissue of *Avena sativa*. Nature 204:286.

Thomas, E.W., B.C. Loughman, and P.G. Powell. 1964b. Metabolic fate of 2,4-dichlorophenoxyacetic acid in the stem tissue of *Phasoolus vulgaris*. Nature 204:884.

Tipton, C.C., R.R. Husted, and F.H.C. Tsao. 1971. Catalysis of simazine hydrolysis by 2,4-dihydroxy-7-methoxy-1,4-benzoxazin-3-one. J. Agr. Food Chem. 19:484.

Tonomura, K., K. Furukawa, and M. Yamada. 1972. Environmental Toxicology of Pesticides. F. Matsumura, G.M. Boush, and T. Misato, editors. Academic Press, 637 pp.

Tsukano, U., and A. Kobayashi. 1972. Formation of γ-BTC in flooded rice field soils treated with γ-BHC. Agr. Biol. Chem. 36:166.

Uesugi, Y., C. Tomizawa, and T. Murai. 1972. Environmental Toxicology of Pesticides. F. Matsumura, G.M. Boush, and T. Misato, editors. Academic Press, 637 pp.

Wain, R.L. 1955. A new approach to selective weed control. Ann. Appl. Biol. 42:151.

Williams, I.H., M.J. Brown, and D.G. Finlayson. 1972. Determination of residues of fensulfothion and its sulfone in muck soil. J. Agr. Food Chem. 20:1219.

Williams, R.T. 1959. Detoxication Mechanisms. Chapman & Hall, London, Second Edition, 796 pp.

Williams, R.T. 1964a. Excerpta Med. Intern. Congr. Ser. 81:9.

Williams, R.T. 1964b. Metabolism of phenolics in animals. In Biochemistry of Phenolic Compounds. J.B. Harborne, editor. Academic Press, p. 205.

Yih, R.Y., D.H. McRae, and H.F. Wilson. 1968. Mechanism of selective action of 3',4'-dichloropropionanilide. Plant Physiol. 43:1291.

Yu, S.J., U. Kiigemagi, and L.C. Terriere. 1971. Oxidative metabolism of aldrin and isodrin by bean root fractions. J. Agr. Food Chem. 19:5.

Zayed, S.M.A.D., I.Y. Mostafa, and A. Hassan. 1965. Metabolism of organophosphorus insecticides. VII. Transformation of ^{32}P-labeled dipterex through microorgansism. Archiv. Eur. Mikrobiologie 51:118.

NONENZYMIC EFFECTS OF PESTICIDES ON MEMBRANES

R.D. O'Brien

Section of Neurobiology and Behavior

Cornell University, Ithaca, New York 14850

The title of this presentation is too general, for I shall only describe our findings on DDT in synthetic membranes. The intent of the DDT work is, however, not only to satisfy our curiosity about DDT's action at the molecular level, but also to explore ways to study nonenzymic effects in membranes for toxicants in general.

The following scheme is rather widely accepted as accounting, at the physiological level, for the action of DDT. It is virtually certain that DDT is a poison of the nervous system, and that most of its action is exerted against the axon. Let me briefly remind you that there are two completely different kinds of transmission in all nervous systems, one which is along axons (long, stringy processes, bundles of which make up the nerve we see running from the brain and spinal cord to our muscles and sense organs); and a completely different process involved in communication between nerves or between a nerve and an effector. The most probable mechanism involved in the transmission of an impulse down an axon is as follows. The axon is normally polarized electrically, with its outside more positive than its inside. This polarization is created by gradients of sodium and potassium ion, which are maintained by metabolic pumps in the axon membrane. These pumps keep the external sodium much higher than the internal sodium, and have the reverse effect upon the potassium concentration. The nerve impulse involves a movement down the axon of a transient permeability to sodium, which permits the sodium to rush in down its concentration gradient, and thereby reduce the polarization of the axon, that is to say to depolarize it. This wave of sodium permeability is chased down the nerve by a corresponding wave of potassium permeability, which permits the potassium to flow out down its concentration gradient, and restore the axon to its original state of polarization. We pictorially describe the transient permeability to sodium as being due to the transient opening of a sodium gate, and of the potassium as being due to the transient opening of a potassium gate. These gates might take the form of actual holes which are briefly opened in the membrane, or they might involve the transient action of some form of carrier mechanism which allows the very polar cation to cross the membrane by jacketing it with a lipid-loving or, as we say, hydrophobic coat. A model for such a hydrophobic jacket will be described later when we talk about valinomycin.

Now let us examine how DDT interferes with this process. There have been a number of studies, but the best and most recent are those of Hille (1958) and Narahashi (1968).

It appears that in both vertebrates and invertebrates, DDT acts by delaying the shutting off of the sodium gate, and also providing some partial block of the potassium gate. The result is to delay the return of the axon to its normal condition after a single impulse has passed by. Because the membrane remains partially depolarized when it should have returned to its normal polarized state, it is relatively easy to excite the nerve again, and consequently, there is a tendency for a stimulus, which in a normal animal would provide only a single impulse to lead to a train of impulses. In an experimental arrangement whereby we can detect nerve impulses by hooking them up to an audio system, then the normal system may sound like this: beep...beep...beep..., but the poisoned animal gives off a beep-beep-beep. This hyperexcitability is seen in the whole insect or vertebrate as a tremoring known as "DDT jitters." Probably death results as a complex series of interferences induced by these jitters, including a failure of communication in control mechanisms, and an exhaustion of reserves of transmitters and energy sources.

These physiological studies made it plausible that the molecular basis of DDT's action was an interference with the gating mechanisms involved in transport of ions across membranes. Our insight into ion transport across membranes was enormously advanced when Mueller and Rudin (1967) described a way in which one could study the ion-transporting properties of synthetic bilayers of phospholipids, and then described the remarkable ability of valinomycin (an antibiotic) to modify it. A sketch of the apparatus which we have used to apply the new understandings to DDT, is shown in Figure 1. Essentially one creates a phospholipid bilayer by painting a solution of lecithin (for instance) plus octane, onto a hole of a fraction of a millimeter diameter which separates two compartments containing electrolyte. Bilayer membranes are thus created, and by inserting electrodes into the electrolyte compartment one can explore the conductivity of these membranes to ions, which can be varied by varying the electrolyte. One finds that unmodified lecithin bilayers have very little permeability to ions of any sort,

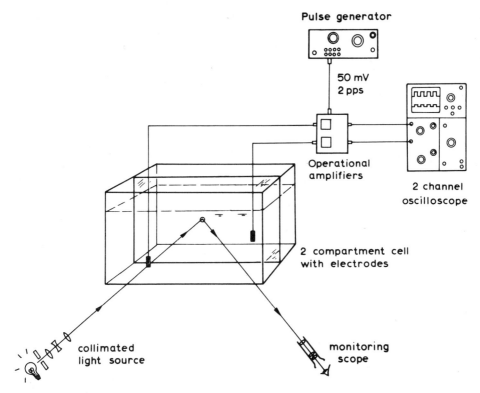

Figure 1.   Cornell version of the Mueller-Rudin apparatus for measuring conductance of synthetic membranes.   The membrane is formed on the small hole in the compartment divider.

which is not surprising by virture of the fact that the membrane interior is made up of a sheet of hydrocarbon tails, creating a kind of lipid blanket which is very hydrophobic, and therefore, very "unsympathetic" to ionic materials. Experimentally, one observes a very low conductance across the membrane, typically 10^{-8} mhos per centimeter squared. If we use a potassium salt as the electrolyte, and now add 10^{-6} M valinomycin, an extraordinary thing happens. The conductance of the membrane is increased up to 1000-fold. This phenomenon, which has been repeated in many laboratories, is almost certainly due to the following. The valinomycin is a donut-shaped molecule with a hydrophobic exterior and a relatively polar interior (Figure 2). It can complex with potassium ions and the complex now is a kind of hydrophobic potassium ion, which can readily move across a lipid layer. Of course there is great interest in the possibility

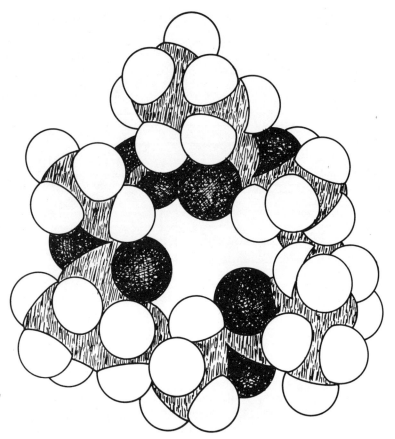

Figure 2.   Valinomycin structure.

that such a mechanism could be involved in the physiological gating
of ions across a membrane, but the question is still entirely open.

We wondered whether, if DDT could interfere with ion gating in
nerves, it might also interfere with gating in this model system.
Hilton and O'Brien (1970) and Hilton et al. (1973) found that it
did. The addition of DDT, at 10^{-6}M, to a system in which potassium
conductance has been induced by valinomycin, resulted in a three-
fold reduction of the potassium conductance. We next proceeded to
explore whether, within a series of active and inactive analogs of
DDT, was there a correlation between their physiological potency
and their ability to affect valinomycin-induced potassium conductance. We were not content to use, as a test of potency, the toxicity towards insects (for instance), because failures of toxicity
might be due to failure to penetrate the nerve, or poor uptake into

the body, or rapid degradation or excretion. Consequently we collaborated with Narahashi's group at Duke University, and studied a group of thirteen analogs with respect to the potency when applied directly to nerves, using either the crayfish or the cockroach. As Table 1 shows, we found a lot of variation in these analogs with respect to their activity on induced conductance. Several of them actually increased the conductance of potassium, and one had absolutely no effect. You should note that it was the relatively polar derivatives of DDT which increased the conductance. There was a very poor correlation between the ability to interfere with induced potassium conductance, and the activity of the analogs on real nerves. However, this finding did not necessarily prove that some similar mechanism might not be operative in the poisoning of the real nerves. For instance, perhaps the order of activity would be different if different phospholipids were used in the synthetic membrane; or perhaps the active agents worked by complexing with the valinomycin, to block formation of a valinomycin-potassium complex, and perhaps the valinomycin was not a good model for the equivalent transport mechanism in real nerves; or perhaps the considerable variation in polarity of the DDT analogs led to a good deal of variation as to how much got into the membrane interior, and the amount which got in might be very different when we were dealing with partitioning between electrolyte and synthetic membrane, as compared with the more complex system in the nerve assay of Narahashi. With respect to the question of varying polarity, Table 1 shows that indeed there were considerable variations. These numbers are based on calculations of partition coefficients by the method of Leo et al (1971). Although there may be minor inaccuracies because these are calculated rather than measured partition coefficients, they do demonstrate the variations to be expected in the extent to which these analogs would partition into lipids from an aqueous system, and the very high partitioning values.

We decided to explore the hypothesis that DDT modified valinomycin's complexation of potassium by forming a DDT-valinomycin complex. To our surprise, all the data was negative. For instance, we were unable to find any change in the spectral characteristics of the valinomycin upon the addition of DDT, and we were unable to demonstrate any effect of DDT upon the so-called bulk transport of potassium by valinomycin. By bulk transport we mean the following. As a consequence of the complexation between valinomycin and potassium, one can show a remarkable effect of valinomycin upon the partitioning of potassium ions between an organic solvent and water. Normally potassium ion scarcely partitions at all out of methylene chloride into water. But if one adds a little valinomycin, a massive migration of potassium into the organic phase can be easily demonstrated. This effect was entirely insensitive to DDT.

How could DDT affect valinomycin-induced conductance, if it did not combine with the valinomycin itself? At about the time we were asking this question, interest was growing about the nature of

TABLE 1

Effects of DDT and Analogs Upon Synthetic Bilayers and Upon Arthropod Nerves

Compound	Effect on Bilayers at 3×10^{-6} M Mean (±SD) ratio Initial:Final Conductance	Effect on Cockroach Sensory Nerves (at 1×10^{-5} M)	Effect on Crayfish Axon (at 1×10^{-4} M)	Log P[a] Calc.
Conductance decreased				
p,p'-DDT	2.8	+	+	6.19
p,p'-fluoro-DDT	2.5	+	+	5.33
p,p'-DDE	2.6	-	-	5.69
o,p'-DDE	1.6	-	-	5.78
Methoxychlor	1.8	+	+	4.93
p,p'-chloro-DDT	1.6			5.80
1-DANP	2.4			5.74
p,p'-DDD	1.7	+	-	5.99
Conductance unaffected				
o,p'-DDD	1.00	+	+	6.08
Conductance increased				
DDMU	0.52	-	-	5.27
p,p'-NH$_2$-DDT	0.48	-	-	2.11
p,p'-NO$_2$-DDT	0.62	+	+	4.27
p,p'-OH-DDT	0.30	-	-	3.41
p,p'-DDA	0.57	-	-	4.86

[a] P = calculated partition coefficient, octanol/water.

the interior of real and synthetic phospholipid membranes. There
was evidence, based particularly upon ESR techniques, that the lipid
interiors were quite fluid, and that the degree of fluidity could
be important in influencing the movement of molecules within the
lipid interior. There was a lot of interest in the possibility that
movements of many things across membranes could involve either shut-
tling, in a valinomycin-like way, or possibly the rotation of proteins (for instance) so that a site on the protein could be exposed
to the exterior of the membrane, could complex with a transported
material, and then rotate to permit the exposure of the transported
material to the inside of the membrane. Whether one was concerned
in translational or rotational transport, one might expect that the
fluidity of the membrane interior could modulate the effectiveness
of transport, and that agents which interfere with transport might
do it by a direct effect upon the fluidity of the interior.

In the case of DDT, this possibility was strengthened by the
observation of Haque et al (1973), that DDT could in fact bind to
lecithin. Their evidence was based upon NMR spectroscopy of mixtures of DDT and lecithin in an organic phase. The binding constants
of these complexes were not very favorable, giving half-maximal
binding at 1.7 M DDT (Haque et al., 1973). However, one can
calculate from the data which I have presented (Table 1) that 10^{-6}
M DDT in an aqueous system would lead to a molar concentration of
well over 1 M in a lipid phase with which the DDT was in equilibrium. And one must remember that it is events within this highly
apolar lipid phase that we are now discussing.

We therefore explored the interaction of DDT and its analogs
with phospholipids, hoping to find (for instance) that DDT could
modify the intermolecular reactions between phospholipids in a way
compatible with a change in membrane fluidity. One approach was
the very classical one of exploring effects of DDT upon monolayers
of phospholipids upon a Langmuir trough. The basic experiment involves adding a known amount of phospholipid to the surface of the
water in the trough. The phospholipid spreads out to form a monolayer, and one then slides the barrier in such a way as to progressively contain the membrane, and with each new containment area
one measures the surface pressure via a torsion balance. At first
the pressure is low, but then as one squeezes the monolayer progressively, the forces increase in a way which is characteristic of the
intermolecular forces between the particular phospholipids, and then
build up to a point where the molecules are squeezed as tightly together as they can, after which further squeezing leads to breaking
of the monolayer just as would happen if one exposed a thin layer of
ice on a pond to an excessive horizontal force. Figure 3 shows a
typical result for a lecithin monolayer. The two points to note
are that there is a relatively long, drawn-out sigmoid relationship,
indicating that intermolecular forces are at work well before the

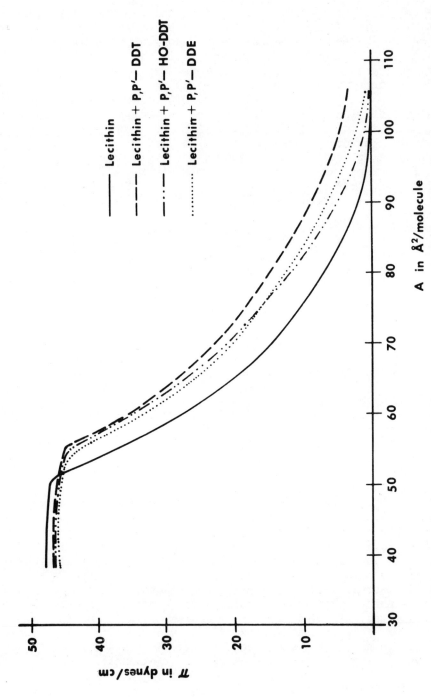

Figure 3. Pressure-area curves of the effect of DDT and two analogs on lecithin monolayers. The Langmuir trough procedure was used.

monolayer is squeezed into its most compacted position. The second
thing to note is that one can measure the surface area per unit of
phospholipid added, by noting the point at which the monolayer breaks.
The figure also shows the effect upon this system of adding DDT and
several analogs. We have data for the whole series on which we
performed the valinomycin work, but only a part of it is shown in
this figure, because of crowding considerations.

These studies led to two conclusions. One was that DDT did
indeed increase the cross-sectional area per phospholipid molecule
by a small amount. We then measured the minimal cross-sectional
area of DDT and its analogs by photographing molecular models.
Our data were compatible with the following hypothesis: the DDT
complexes with lecithin to form a 1:1 complex, and the complex now
has the minimal cross-sectional area of the DDT, which is somewhat
larger than the lecithin. The same appears to hold true for all
the analogs, and consequently the modest variations in their effect
upon the monolayer could be accounted for entirely by variations in
their cross-sectional area. But there was no indication that physiologically active compounds behaved in a way different from the
physiologically inactive ones. Furthermore, it was clear that the
intermolecular forces, as shown by the sigmoidal shape of the force-
area curves, were not significantly modified by any of the compounds.
When we explored cholesterol monolayers, Figure 4 shows that they
were completely unaffected by DDT, and that the shape of the curves
was drastically different, indicating very little intermolecular
interactions until the film was squeezed very close together. When
we looked at phospholipid monolayers made from extracts of electroplax membrane of electric skate, which is rich in neural membranes,
we found an effect like that with lecithin (Hilton and O'Brien,
1973).

These observations confirm, by a drastically different technique, the observations of Haque and Tinsley that complexes between
DDT and lecithin could occur. But they did not reveal anything interesting about the consequences of such complex formation for phospholipid interaction. However, one must note that in a monolayer,
the DDT is free to make complexes with the hydrophobic component
of the phospholipids which is exposed to air. In a bilayer, by
constrast, the lipid side of the monolayer from the inside surface
of the membrane is juxtaposed tightly, and doubtless mingles with,
the lipid layer from the outermost phospholipid surface. Consequently, it is possible that DDT could have only modest effects
upon a monolayer, and much more severe effects upon the more complex
lipid interaction involved in a bilayer.

At this point we began collaborations on a new technique developed by Professors Webb and Elson at Cornell, the experiments which
I shall describe being performed by Dr. Tom Herbert. Let me begin

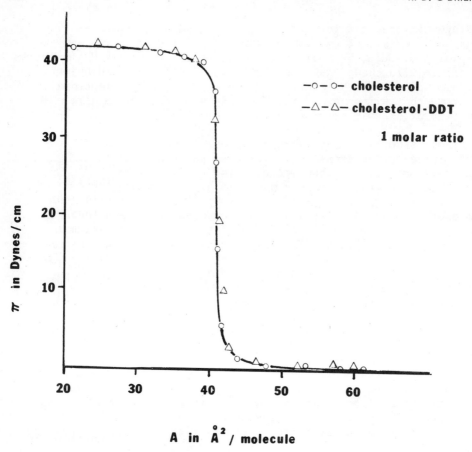

Figure 4.   Lack of effect of DDT on pressure-area curve for
cholesterol.

by describing the principles and practice of the basic experiment,
and then go on to the DDT effect.   The technique, which is called
fluorescent correlation spectroscopy, provides a way to measure the
rates of movement of fluorescent probes in various environments, in
this case within a membrane.   Suppose we imagine illuminating a very
small volume element within a membrane, within which there is a
fluorescent material.   If the illuminated volume is small enough,
then the effects of individual fluorescent molecules drifting into
and out of that volume because of thermal agitation can be quite
large, and one therefore would see a fluctuation in the fluorescent
intensity.   In practice, an area in the film about 10 microns in
diameter is illuminated with a laser beam, and the fluorescent output is collected in a parabolic reflector, which is directed upon

a photomultiplier, which is connected to an online computer which
calculates a function called the auto-correlation coefficient.
This is a parameter which measures the extent of fluctuation of
the fluorescent output. As you can imagine, if one reduces the
temperature of the whole system, the rate of movement in and out
is reduced, therefore the fluctuation is reduced. Similarly, if
any agent reduces the fluidity of the lipid interior within which
all these events occur, the rate of fluctuation will be reduced
and the time characteristics of the auto-correlation coefficient
will be affected appropriately. We used this technique to study
the fluidity of membranes made of phospholipid extracted from
Torpedo electroplax as described above. The first point to make is
that the method was indeed successful, and this is no small task,
because one has to detect a very small fluctuation in signal against
a rather substantial background of noise; the key to this process is
to gather rather a lot of data for analysis, and each run takes
about six hours. Values were thereby obtained for the diffusion
coefficient of the probe which was used, which chemically was di
(octadecyl)oxycarbocyanin, and therefore involved two long alkyl
chains attached to a cyclic fluorescent head. The diffusion coefficient was 0.63 cm^2/sec. Experiments were conducted with DDT and
with a physiologically inactive analog, hydroxy DDT. The next point
to make is a disappointing one. DDT had absolutely no effect upon
the fluidity of the interior of these membranes, as judged by the
diffusion coefficient of the fluorescent probe. But an entirely
unexpected phenomenon was observed. DDT at a concentration of 3 x
10^{-6}M displaced almost half of the fluorescent probe from the membrane.

Whether this interesting phenomenon has physiological importance is at the moment a matter of pure speculation; but let me
speculate. We have here a kind of physical analog of competitive
inhibition in enzyme systems. Can it be that the ability of one
compound to displace another from a membrane is a function purely
of its partition coefficient? But Table 1 shows that there is no
simple correlation between partitioning ability of DDT analogs and
their effects upon valinomycin-induced conduction. It seems very
likely that, in addition to the simple partitioning power which is
measured by the partition coefficient, the ability of the molecule
to interact with the components of the lipid system should also
play a role, so that compounds which were equal in partitioning
properties might differ very greatly in their ability to push out
another molecule. All this is entirely new thinking, and yet one
can readily conceive of experimental approaches to measuring the
relative displacing abilities of various small molecules within
various phospholipid membranes, and we are undertaking such experiments now. The possibility that I am leading up to is that DDT may
interfere with ion transport in membranes, not by a simple change

in the fluidity of the interior, but rather by competitive displacement or interference with molecules involved in the transport process. If this turns out to be the case, we may find that DDT is a valuable tool to explore the transport process, and it may open new windows onto the question of the ways in which modulation of transport processes can be explained.

ACKNOWLEDGMENT

This work was supported in part by Hatch funds and National Institutes of Health Grant ES 00901.

REFERENCES

Haque, R., I.J. Tinsley, and D.W. Schmedding. 1973. Lipid binding and mode of action of compounds of the dichlorodiphenyltrichloroethane type: A proton magnetic resonance study. Mol. Pharmacol. 9:17.

Hille, B. 1958. Charges and potentials of the nerve surface divalent ions and pH. J. Gen. Physiol. 51:221.

Hilton, B.D., T.A. Bratkowski, M. Yamada, T. Narahashi, and R.D. O'Brien. 1973. The effects of DDT analogs upon potassium conductance in synthetic membranes. Pestic. Biochem. Physiol. 3:14.

Hilton, B.D., and R.D. O'Brien. 1970. Antagonism by DDT of the effect of valinomycin on a synthetic membrane. Science 168:841.

Hilton, B.D., and R.D. O'Brien. 1973. The effects of DDT and its analogs upon lecithin and other monolayers. Pestic. Biochem. Physiol. 3:206.

Leo, A., C. Hansch, and D. Elkins. 1971. Partition coefficients and their uses. Chem. Rev. 71:525.

Mueller, P., and R. Rudin. 1967. Development of K^+-Na^+ discrimination in experimental bimolecular lipid membranes by macromolecular antibiotics. Biochem. Biophys. Res. Comm. 26:398.

Narahashi, T., and H.G. Haas. 1968. Interaction of DDT with the components of lobster nerve membrane conductance. J. Gen. Physiol. 51:177.

PHYSIOLOGICAL EFFECTS OF CHLORINATED HYDROCARBONS ON AVIAN SPECIES

David B. Peakall

Section of Ecology and Systematics

Cornell University, Ithaca, New York 14850

The persistent organochlorines have been shown to effect a number of enzyme systems in mammals and birds. Many of these effects are found only at high concentrations, and it is proposed here to concentrate on those effects which have or may have relevance to existing environmental levels.

The best documented case of an organochlorine, specifically DDT and its metabolites, causing an effect in the environment is eggshell thinning in predatory birds. Although the case is well documented, it has not been without its critics. Hazeltine (1972) criticized the dose-response data for DDE against eggshell thinning for the Brown Pelican. The sample of eggs examined by Hazeltine was only nine, and his data of increasing thickness with increasing DDE concentration has to be set against the comprehensive studies of Blus et al. (1972a, b) and Risebrough (1971). These workers, with a total sample size of 279 eggs, found a highly significant negative correlation. Detailed studies of the Peregrine in Great Britain (Ratcliffe, 1970) and North America (Cade et al., 1971; Peakall et al., in press) have shown an essentially linear relation between the log of concentration of DDE against degree of eggshell thinning. In Figure 1, the log of the mean total DDT concentration is plotted against the average eggshell thinning for 21 studies by 14 different observers of 9 species of birds of prey. It should be made clear that the DDT in the egg does not cause the eggshell thinning. The critical level is considered to be the level in the oviduct (see later); the level in the egg is only an expression of the level in the female.

Another criticism that has been leveled is that eggshell thinning occurred almost immediately after the first usage of DDT

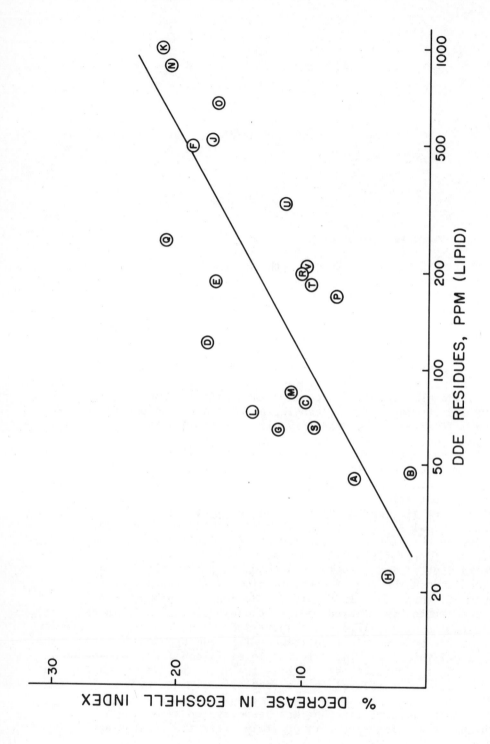

Figure 1. Relationship between degree of eggshell thinning and DDE residue levels for various species of Falconiformes. Each point represents the mean of degree of thinning plotted against mean of DDE residues. In some cases, the shell thickness measurements were carried out on a larger sample than the sample analyzed for residues. Correlation coefficient is 0.73.

A-D	Cooper's Hawk (*Accipiter cooperii*)	Snyder et al., 1973
E	European Sparrow Hawk (*A. nisus*)	Newton, 1973a
F		Koeman et al., 1972
G	Redtailed Hawk (*Buteo jamaicensis*)	Seidensticker and Reynolds, 1972
H	Rough-legged Hawk (*B. lagopus*)	Cade et al., 1971
J,K	Bald Eagle (*Haliaeetus leucocephalus*)	Postupalsky, 1971
L	Prairie Falcon (*Falco mexicanus*)	Enderson and Berger, 1970
M		Fyfe et al., 1969
N-P	Peregrine (*F. peregrinus*)	Cade et al., 1971
Q		Berger et al., 1970
R,S		Ratcliffe, 1970
T	Merlin (*F. columbarius*)	Temple, 1972
U		Newton, 1973b
V	American Kestrel (*F. sparverius*)	Lincer, 1972.

and before its use became widespread (Gunn, 1972). The shape of
the dose-response curve gives an explanation for this phenomenon;
only a small amount of DDT is needed to cause a profound effect.
Direct experimental evidence of the presence of DDE in Peregrine
eggs laid during this period of the 1940's has recently been obtained by elution of the eggshell membranes (Peakall, 1974; and
unpublished data). The rapid occurrence of DDE after the first
introduction of DDT and the simultaneous occurrence of eggshell
thinning have been demonstrated from eggs collected in Great
Britain (Figure 2).

Although pesticide-induced eggshell thinning has been clearly
demonstrated in some species, it should be pointed out that there
is a wide range of response for different species. The available
data is summarized in Table 1. Two major points can be made from
this compilation. First that there is a wide range of response;
e.g., 10 ppm DDE in the diet of American Kestrels causes nearly
30% thinning (Peakall et al., 1973) whereas, 300 ppm causes no
significant thinning in chickens (Lillie et al., 1972). Therefore,
it would be expected that eggshell thinning could become critical
in some species but have no significant effect in other species.
The second point is that studies so far indicate that species high
in the trophic levels are those which are the most sensitive. The
hazard to Peregrines and Brown Pelicans is high because: a) they
are sensitive to DDE-induced eggshell thinning; and b) they are
high in the food chain. Conversely, for quail the hazard is low
because: a) they are low in the food chain; and b) they have low
sensitivity to DDE-induced eggshell thinning.

The finding that eggshell thinning may be as much as 95% in
extreme circumstances in the Brown Pelican off California (Risebrough et al., 1971) shows that profound effects can occur under
field conditions. Further, it argues strongly that the major site
of action is in the oviduct itself. Generalized hypocalcemia of
this order of magnitude would be incompatible with normal function.
Experimental studies have shown no alteration of blood calcium
levels caused by DDE under conditions causing significant eggshell
thinning (Peakall and Kinter, unpublished). It seems reasonable
to ignore, for the moment, such systems as vitamin D and other
steroid hormones and the parathyroid hormone and to concentrate
on calcium transport in the oviduct itself and the actual process
of eggshell formation. An outline of these processes are given in
Figure 3.

The oviduct itself stores very little calcium, and thus calcium must be transported from the blood to the site of shell formation continuously during the process of shell formation. Both
ATPases and calcium binding protein are involved. Alkaline phosphatase has been shown to be involved in the uptake of calcium from
the intestine, probably coupled with ATPase, but this does not

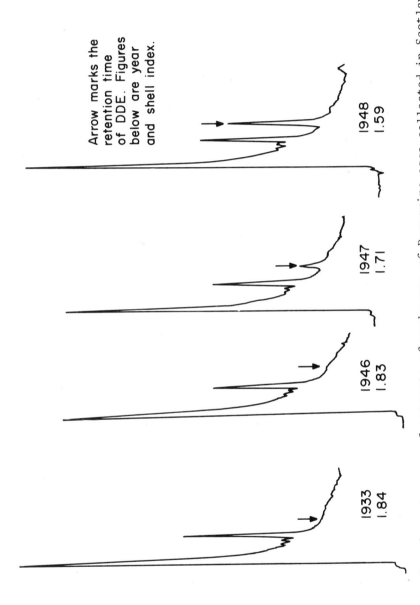

Figure 2. Gas chromatograms of extracts of membranes of Peregrine eggs collected in Scotland 1933 to 1948. The shell index is defined as weight (mg) divided by the product of the length times breadth (mm); shell index is closely correlated with eggshell thickness (Ratcliffe, 1970). Unpublished data using procedure described in Peakall (1974).

TABLE 1

Species Response to DDE-Induced Eggshell Thinning

	Order	Species	Reference
Highly sensitive	Pelecaniformes	Brown Pelican (*Pelecanus occidentalis*)	Blus et al., 1972a,b Risebrough, 1971
		Double-crested Cormorant (*Phalacrocorax auritus*)	Risebrough et al., 1971 Anderson et al., 1969 Postupalsky, 1971
	Ciconiiformes	Several species	Faber and Hickey, 1973
	Falconiformes	Several species	See Figure 1
	Strigiformes	Schreech Owl (*Otus asio*)	McLane and Hall, 1972
Moderately sensitive	Anseriformes	Mallard (*Anas platyrnynchos*)	Heath et al., 1969 Peakall et al., 1973
		Black Duck (*Anas rubripes*)	Longcore et al., 1971
	Charadriiformes	Herring Gull (*Larus argentatus*)	Hickey and Anderson, 1968
	Columbiformes	Ring Dove (*Streptopelia risoria*)	Peakall, 1970 Peakall et al., 1973
Low sensitivity	Galliformes	Quail (*Corturnix coturnix*) (*Colinus virginianus*)	Bitman et al., 1969 Heath et al., 1972
		Chicken (*Gallus gallus*)	Davison and Sell, 1972 Cecil et al., 1973 Lillie et al., 1972
	Passeriformes	Bengalese Finch (*Lonchura striata*)	Jefferies, 1969

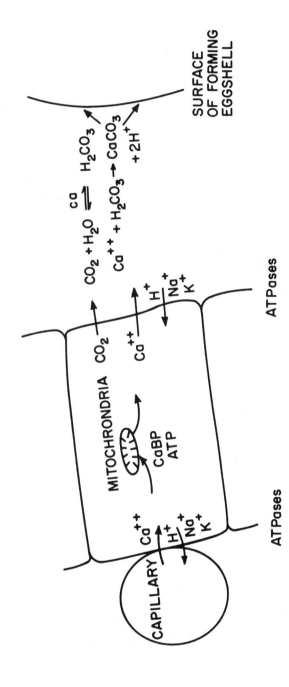

Figure 3. Diagrammatic representation of the processes involved in calcium transport and eggshell formation in the avian oviduct. After Schraer and Schraer (1970) and Schraer et al. (1973).

appear to have been studied in the oviduct. In the lumen of the
shell gland, carbon dioxide is hydrated by the enzyme carbonic
anhydrase to form carbonic acid. The source of the carbon dioxide
is considered to be metabolic from the shell gland rather than
transported (Lorcher and Hodges, 1969). The carbonic acid then
reacts with calcium to form the insoluble calcium carbonate which
precipitates to form the eggshell.

Decreased activity of carbonic anhydrase in the oviduct following the administration of DDT was reported simultaneously by
Bitman et al. (1970) and Peakall (1970). Although some early
studies (Torda and Wolff, 1949 and Keller, 1952) claimed *in vitro*
inhibition of carbonic anhydrase by DDT, this point is disputed by
more recent work. Dvorchik et al. (1971), using human red cells,
failed to find inhibition with DDT or DDE. A more detailed study
by Pocker et al. (1971) also failed to find inhibition of carbonic
anhydrase from bovine erthrocytes and spinach with DDT, DDE, or
dieldrin. Both groups found co-precipitation at higher concentrations of the organochlorines. Pocker and co-workers concluded that
when considering the effect of DDT, DDE, and dieldrin on eggshell
formation, a purely chemical inhibition of carbonic anhydrase could
be definitely ruled out. Recent studies in my laboratory suggest
that their extrapolation from *in vitro* experiments to an *in vivo*
phenomenon is not correct.

Initially, the dose response of dietary DDE and dietary acetazolamide (a known carbonic anhydrase inhibitor) were studied.
The results are shown in Figure 4. The plot of shell thickness
against acetazolamide concentration clearly shows the rapid initial
decrease typical of DDE-induced eggshell thinning. The decrease in
eggshell thickness is paralleled by a decrease in carbonic anhydrase activity. Above 100 ppm acetazolamide, no further decrease
in eggshell thickness occurs and carbonic anhydrase activity in the
shell gland homogenate is reduced to zero. This is the maximum
effect of carbonic anhydrase inhibition on eggshell thickness. The
effect of DDE is similar in its initial response, but the definite
plateau effect is not observed. Further, the reduction in carbonic
anhydrase activity is not as great for the same degree of eggshell
thinning with DDE as occurs with acetazolamide. Experiments with
labeled acetazolamide and labeled DDE show a clear difference in
the cellular distribution of these two compounds. The labeled
acetazolamide is found largely in the supernatant while the DDE is
associated with membrane fractions.

The inhibition of carbonic anhydrase by DDE in the oviduct
membrane, in a manner parallel to the findings of Enns (1972) in
the red blood cell, would reduce the supply of H_2CO_3 available for
shell formation.

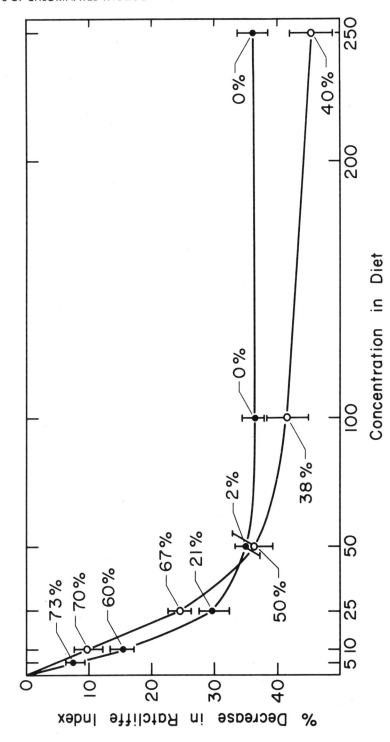

Figure 4. Eggshell thinning in Ring Doves exposed to diets containing DDE or acetazolamide. Acetazolamide = 0; DDE = 0. The carbonic anhydrase activity is expressed as percentage of control. CaBP = calcium binding protein; ca = carbonic anhydrase; ATP = adenosine triphosphate.

The failure of *in vitro* experiments to show inhibition of carbonic anhydrase by DDE appears to be due to the organic solvent, usually dimethyl formamide, necessary to get the DDE into solution. Addition of DMF to tissue homogenates of the oviducts of DDE-exposed birds restores the carbonic anhydrase activity to control values. Current findings from both *in vivo* (Bitman et al., 1970; Peakall, 1970; and unpublished data) and *in vitro* (Dvorchik et al., 1971; Pocker et al., 1971) experiments can be explained by a weak binding of DDT and DDE to carbonic anhydrase in the membrane causing decreased activity of the enzyme and the breaking of this bond in the presence of organic solvents thus restoring the activity of the enzyme.

The difference between the dose response of DDE and acetazolamide on eggshell thickness indicates that a second mechanism is involved at high concentrations. Preliminary work (Pritchard et al., 1972) suggests that inhibition of ATPases may be involved. While the effect of chlorinated hydrocarbons on the several cation-activated ATPases in the oviduct is limited, there is considerable data from other systems. Most studies of inhibition caused by chlorinated hydrocarbons are from *in vitro* mammalian preparations. Studies, both *in vivo* and using isolated sacs, have shown an effect of DDT on osmoregulation and on ATPase activity in eels and killifish (Janicki and Kinter, 1971; Kinter et al., 1972).

No effect was found on the activity of calcium-binding protein in the oviduct of ducks treated with DDE, nor in *in vitro* experiments (Pritchard et al., 1972).

To conclude on eggshell thinning, it is considered that the reality of DDT-induced eggshell thinning has been established for some, but not all species. For some predatory species in some areas, the effect of eggshell thinning has seriously impaired reproduction. The major components of the mechanism of eggshell thinning appear to be localized in the enzyme systems of the oviduct itself.

There has been a vast amount of research on the effects of organochlorines on hepatic microsomal enzyme induction (see reviews Conney, 1967; Conney and Burns, 1972). The bulk of this work has been done using *in vitro* mammalian systems. The minimum effective dose of organochlorines to cause an alteration of liver enzyme activity in mammals is 1-3 ppm in the diet (Gillett et al., 1966; Gillett, 1968; Hoffman et al., 1970; Hodge et al., 1967). Studies on birds are less complete and a good deal more inconsistent.

Gillett et al. (1966) found a two- to three-fold increase in the specific activity of aldrin and heptachlor epoxidation in female quail fed 100 ppm DDT for two weeks; whereas, males showed inhibition to levels of about 60% of controls. The picture is further complicated by the fact that the control levels were 5-10 times higher in males than in females.

Gillett and Arscott (1969) found that 5 ppm dieldrin in the diet of young quail showed increased hepatic microsomal epoxidase activity an no sex-linked differences were found. The same workers found that 100 ppm DDT in the diet for 27 weeks caused decreased epoxidase activity, in contrast to previous findings of induction after two weeks exposure.

Abou-Donia and Menzel (1968) showed that DDT, DDD, and DDE all stimulated hepatic microsomal dealkylation in chicks whose eggs had been treated with these compounds.

Peakall (1967) showed that DDT and dieldrin induced hepatic enzymes in male pigeons that increased the rate of metabolism of progesterone and testosterone. Risebrough et al. (1968) extended this work to cover DDE and PCB in the metabolism of oestradiol in pigeons. Increased metabolism of oestradiol in microsomal extracts from Kestrels after exposure to PCBs was shown by Lincer and Peakall (1970).

Lustick and Peterle (1971) found that 5 ppm DDT in the diet of Bobwhite Quail caused a doubling of the rate of conversion of progesterone. In males, these workers found that the conversion of testosterone to more polar metabolites was inversely related to testes size.

Sell et al. (1971) found that 100 and 200 ppm DDT caused a significant decrease of hydroxylase activity (50% of controls) and N-demethylase activity (70-75% of controls) in chickens exposed for 12 weeks. Dieldrin at 20 ppm caused slight reduction of hydroxylase activity and significant induction of N-demethylase activity. Stephen et al. (1970) were unable to demonstrate enzyme induction in chickens (either chicks or adult hens) using a single i.p. injection of 10 mg/kg p,p'-DDT for chicks or 20 mg/kg for hens.

Bitman et al. (1971) found that both DDT and DDE at a diet dose of 100 ppm increased the sleeping time due to phenobarbital in quail, indicating decreased hepatic activity. While direct studies on enzyme induction were not made, previous work on rats has shown a parallel between decreased sleeping time and hepatic enzyme induction (Hart and Fouts, 1965).

Nowicki and Norman (1972) gave chicks p,p'-DDT, o,p'-DDT, Aroclor 1254 and Aroclor 1260 at 1 mg/kg for seven days. These workers found that using an *in vitro* incubation system that the metabolism of testosterone, estradiol -17B, and androstene -3,17- dione by a hepatic fraction was increased three-fold by all the organochlorines studied.

There is more variation between species and between dosage and time of exposure among birds than among mammals. High doses

(100 or more ppm in diet) for prolonged periods appear to favor inhibition, and this may be due to sub-lethal toxic effects. Apparently, no reasons are known for this variation; but any work on birds should include as little species-to-species and dose-to-dose projection as possible.

Despite all the studies on enhanced metabolism of steroids, very little information is available on the physiological consequences. Assuming that the rate of steroid metabolism in the liver is increased *in vivo* by organochlorines, this would lead to a decreased level of circulating levels only if the increased metabolism was not compensated for by increased production. Alterations of hormone levels could cause changes in reproductive behavior, and the occurrence of secondary sex characters to be depressed or delayed. Delay in egg-laying caused by DDT or DDE has been noted under experimental conditions in the Bengalese Finch (Jeffries, 1967) and Ring Dove (Peakall, 1970; Haegele and Hudson, 1973). In the field, lack of nest site tenacity in Brown Pelican (Schreiber, unpublished), deliberate egg-breaking by Herons (Milstein et al., 1970), and premature pupping of Sea Lions (DeLong et al., 1973) has been blamed on organochlorines with suggestions that hormone imbalance could be exposed to enough organochlorines to cause enzyme induction, but definitive experiments remain to be carried out.

One laboratory study (Peakall, 1970) decreased blood estradiol levels and increased hepatic levels were demonstrated at the midway point between paring and egg laying. More detailed studies are needed to correlate field findings with physiological and biochemical alterations.

SUMMARY

The persistent organochlorine pesticides have been widely considered to have a marked detrimental effect on some avian populations. Aside from occasional instances, such as poisoning via seed dressings, most effects have been on reproduction rather than direct mortality.

Eggshell thinning, caused by DDT and its metabolites, has been demonstrated in some, but not all, species. In some areas, predatory species have accumulated enough DDT to exert a marked physiological effect. The mechanism of eggshell thinning is considered to be due to inhibition of carbonic anhydrase and ATPase in the oviduct.

Many organochlorines have been found to cause hepatic enzyme induction, and *in vitro* preparations cause increased metabolism of steroids. Behavioral effects during the reproductive cycle have been claimed but definitive evidence is lacking.

REFERENCES

Abou-Donia, M.B., and D.B. Menzel. 1968. Chick microsomal oxi-
    dases. Isolation, properties, and stimulation by embryonic
    exposure to 1,1,1-trichloro-2,2-bis(p-chlorophenyl)ethane.
    Biochem. 7:3788-3794.

Anderson, D.W., J.J. Hickey, R.W. Risebrough, D.F. Hughes, and
    R.E. Christensen. 1969. Significance of chlorinated hydro-
    carbon residues to breeding pelicans and cormorants. Can.
    Field Nat. 83:91-112.

Berger, D.D., D.W. Anderson, J.D. Weaver, and R.W. Risebrough.
    1970. Shell thinning in eggs of Ungava Peregrines. Can.
    Field Nat. 84:265-267.

Bitman, J., H.C. Cecil, and G.F. Fries. 1970. DDT-induced inhi-
    bition of avian shell gland carbonic anhydrase: a mechanism
    for thin eggshells. Science 168:594-596.

Bitman, J., H.C. Cecil, S.J. Harris, and G.F. Fries. 1969. DDT
    Induces a decrease in eggshell calcium. Nature 224:44-46.

Bitman, J., H.C. Cecil, S.J. Harris, and G.F. Fries. 1971. Com-
    parison of DDT effect on pentobarbital metabolism in rats and
    quail. J. Agric. Food Chem. 19:333-338.

Blus, L.J., C.D. Gish, A.A. Belisle, and R.M. Prouty. 1972a. Loga-
    rithmic relationship of DDE residues to eggshell thinning.
    Nature 235:376-377.

Blus, L.J., C.D. Gish, A.A. Belisle, and R.M. Prouty. 1972b. Fur-
    ther analysis of the logarithmic relationship of DDE residues
    to eggshell thinning. Nature 240:164-166.

Cade, T.J., J.L. Lincer, C.M. White, D.G. Roseneau, and L.G. Swartz.
    1971. DDE residues in eggshell changes in Alaskan falcons and
    hawks. Science 172:955-957.

Cecil, H.C., J. Bitman, G.F. Fries, S.J. Harris, and R.J. Lillie.
    1973. Changes in eggshell quality and pesticide content of
    laying hens or pullets fed DDT in high or low calcium diets.
    Poult. Sci. 52:648-653.

Conney, A.H. 1967. Pharmacological implications of microsomal
    enzyme induction. Pharmacol. Rev. 19:317-366.

Conney, A.H., and J.J. Burns. 1972. Metabolic interactions among
    environmental chemicals and drugs. Science 178:576-586.

Davison, K.L., and J.L. Sell. 1972. Dieldrin and p,p'-DDT effect
    on egg production and eggshell thickness of chickens. Bull.
    Environ. Contam. Toxicol. 7:9-18.

DeLong, R.L., W.G. Gilmartin, and J.G. Simpson. 1973. Premature
    births in California Sea Lions: association with high organo-
    chlorine pollutant residue levels. Science 181:1168-1170.

Dvorchik, B.H., M. Istin, and T.H. Maren. 1971. Does DDT inhibit
    carbonic anhydrase? Science 172:728-729.

Enderson, J.H., and D.D. Berger. 1970. Pesticides: eggshell
    thinning and lowered production of young in Prairie Falcons.
    BioScience 20:355-356.

Enns, T. 1972. Carbonic anhydrase, red cell membranes, and CO_2
    transport. Adv. Exp. Med. Biol. 28:201-208.

Faber, R.A., and J.J. Hickey. 1973. Eggshell thinning, chlorinated
    hydrocarbons, and mercury in inland aquatic bird eggs, 1969 and
    1970. Pestic. Monit. J. 7:27-36.

Fyfe, R.W., J. Campbell, B. Hayson, and K. Hodson. 1969. Regional
    population declines and organochlorine insecticides in Canadian
    Prairie Falcons. Can. Field Nat. 83:191-200.

Gillett, J.W. 1968. "No effect" level of DDT in induction of
    microsomal epoxidation. J. Agric. Food Chem. 16:295-297.

Gillett, J.W., and G.H. Arscott. 1969. Microsomal epoxidation in
    Japanese Quail. Induction by dietary dieldrin. Comp. Biochem.
    Physiol. 30:589-600.

Gillett, J.W., T.H. Chan, and L.C. Terriere. 1966. Interactions
    between DDT analogs and microsomal epoxidase systems. J.
    Agric. Food Chem. 14:540-545.

Gunn, D.L. 1972. Dilemmas in conservation for applied biologists.
    Ann. Appl. Biol. 72:105-127.

Haegele, M.A., and R.H. Hudson. 1973. DDE effects on reproduction
    of Ring Dove. Environ. Pollut. 4:53-58.

Hart, L.G., and J.R. Fouts. 1965. Further studies on the stimula-
    tion of hepatic microsomal drug metabolizing enzymes by DDT
    and its analogs. Arch. Exp. Path. Pharmacol. 249:486-500.

Hazeltine, W. 1972. Disagreements on why Brown Pelican eggs are
    thin. Nature 239:410-411.

Heath, R.G., J.W. Spann, and J.F. Kreitzer. 1969. Marked DDE
    impairment of Mallard reproduction in controlled studies.
    Nature 224:47-48.

Heath, R.G., J.W. Spann, J.F. Kreitzer, and C. Vance. 1972. Ef-
    fects of polychlorinated biphenyls on birds. 13th Int.
    Orinthol. Congr. 475-485.

Hickey, J.J., and D.W. Anderson. 1968. Chlorinated hydrocarbons
    and eggshell changes in raptorial and fish-eating birds.
    Science 162:271-273.

Hodge, H.C., A.M. Boyce, W.B. Deichmann, and H.F. Kraybill. 1967.
    Toxicology and no-effect levels of aldrin and dieldrin.
    Toxicol. Appl. Pharmacol. 10:613-675.

Hoffman, D.G., H.M. Worth, J.L. Emmerson, and R.C. Anderson. 1970.
    Stimulation of hepatic drug-metabolisming enzymes by DDT; the
    relationship of liver enlargement and hepatotoxicity in the
    rat. Toxicol. Appl. Pharmacol. 16:171-178.

Janicki, R.H., and W.B. Kinter. 1971. DDT: disrupted osmoregu-
    latory events in the intestine of the eel *Anguilla rostrata*
    adapted to seawater. Science 173:1146-1148.

Jefferies, D.J. 1969. Induction of apparent hyperthyroidism in
    birds fed DDT. Nature 222:578-579.

Jefferies, D.J. 1967. The delay in ovulation produced by p,p'-
    DDT and its possible significance in the field. Ibis 109:
    266-272.

Keller, H. 1952. Cited in Methods of Enzymatic Analysis. H.V.
    Bergmeyer Ed. Acad. Press, 1965.

Kinter, W.B., L.S. Merkens, R.H. Janicki, and A.M. Guarino. 1972.
    Studies on the mechanism of toxicity of DDT and polychlorinated
    biphenyls: disruption of osmoregulation in marine fish.
    Environ. Health Perspect. 1:169-173.

Koemann, J.H., C.F. van Beusekom, J.J.M. deGoeij. 1972. Eggshell
    and population changes in the Sparrow-hawk (*Accipiter nisus*)
    TNO-Nieuws 27:542-550.

Lillie, R.J., C.A. Denton, H.C. Cecil, J. Bitman, and G.F. Fries.
    1972. Effect of p,p'-DDT, o,p'-DDT, and p,p'-DDE on the re-
    productive performance of caged White Leghorns. Poult. Sci.
    51:122-129.

Lincer, J.L. 1972. The effects of organochlorines on the American Kestrel (*Falco sparverius* Linn). Ph.D. Thesis, Cornell Univ.

Longcore, J.R., F.B. Samson, and T.W. Whittendale, Jr. 1971. DDE thins eggshells and lowers reproductive success of captive Black Duck. Bull. Environ. Contam. Toxicol. 6:485-490.

Lorcher, K., and R.D. Hodges. 1969. Some possible mechanisms of formation of the carbonate fraction of eggshell calcium carbonate. Comp. Biochem. Physiol. 28:119-128.

Lustick, S., and T.J. Peterle. 1971. The effect of feeding low levels of DDT on the steroid metabolism in Bobwhite (*Colinus virginianus*). 89th Meeting, Am. Ornithol. Union, Seattle.

McLane, M.A.R., and L.C. Hall. 1972. DDE thins Schreech Owl eggshells. Bull. Environ. Contam. Toxicol. 8:65-68.

Milstein, P.L.S., I. Presst, and A.A. Bell. 1970. The breeding cycle of the Grey Heron. Ardea. 58:171-257.

Newton, I. 1973a. Success of Sparrowhawks in an area of pesticide usage. Bird Study 20:1-8.

Newton, I. 1973b. Egg breakage and breeding failure in British Merlins. Bird Study 20:241-244.

Nowicki, H.G., and A.W. Norman. 1972. Enhanced hepatic metabolism of testosterone, 4-androstene-3,17-dione, and estradiol-17B in chickens pretreated with DDT or PCB. Steroids 19:85-99.

Peakall, D.B. 1967. Pesticide-induced enzymatic breakdown of steroids in birds. Nature 216:505-506.

Peakall, D.B. 1970. p,p'-DDT: effect on calcium metabolism and concentration of estradiol in the blood. Science 168:592-594.

Peakall, D.B. 1974. DDE: its presence in Peregrine eggs in 1948. Science 183:673-674.

Peakall, D.B., T.J. Cade, C.M. White, and J.R. Haugh. Organochlorine residues in Alaskan Peregrines. Pestic. Monit. J. (in press).

Peakall, D.B., J.L. Lincer, R.W. Risebrough, J.B. Pritchard, and W.B. Kinter. 1973. DDE-induced eggshell thinning: structural and physiological effects in three species. Comp. Gen. Pharmacol. 4:305-311.

Pocker, Y., M.W. Bueg, and V.R. Ainardi. 1971. Coprecipitation of carbonic anhydrase by 1,1-bis(p-chlorophenyl)-2,2,2-trichloroethane, 1,1-bis(p-chlorophenyl)-2,2,-dichroethylene, and dieldrin. Biochem. 10:1390-1396.

Postupalsky, S. 1971. Toxic chemicals and declining Bald Eagle and Cormorants in Ontario. Can. Wildl. Serv. Ms. Rep. No. 20.

Pritchard, J.B., D.B. Peakall, R.W. Risebrough, and W.B. Kinter. 1972. DDE-induced eggshell thinning in White Pekin Ducks, *Anas platyrynchos*: structural, physiological, and biochemical studies. Bull. Mt. Desert Isl. Biol. Lab. 12:77-79.

Ratcliffe, D.A. 1970. Changes attributable to pesticides in egg breakage frequency and egghsell thickness in some birds. J. Appl Ecol. 7:67-116.

Risebrough, R.W. 1971. Effects of environmental pollutants upon animals other than man. 6th Berkeley Symp. on mathematical statistics and probability.

Risebrough, R.W., P. Reiche, D.B. Peakall, S.G. Herman, and M.N. Kirven. 1968. Polychlorinated biphenyls in the global ecosystem. Nature 220:1098-1102.

Risebrough, R.W., F.C. Sibley, and M.N. Kirven. 1971. Reproductive failure of the Brown Pelican on Anacapa Island in 1969. Am. Birds 25:8-9.

Schraer, R., and G. Schraer. 1971. The avian shell gland: a study in calcium transport. Biological Calcification. Appleton-Century-Croft.

Schraer, R., J.A. Elder, and H. Schraer. 1973. Aspects of mitochondrial function in calcium movement and calcification. Fed. Proc. 32:1938-1943.

Seidensticker, IV, J.C., and H.V. Reynolds, III. 1971. The nesting, reproductive performance, and chlorinated hydrocarbon residues in the Redtailed Hawk and Great Horned Owl in south-central Montana. Wilson Bull. 83:408-418.

Sell, J.L., K.L. Davison, and R.L. Puyear. 1971. Aniline hydroxylase, N-demethylase, and cytochrome P_{450} in liver microsomes of hens fed DDT and dieldrin. J. Agric. Food Chem. 19:58-60.

Snyder, N.F.R., H.A. Snyder, J.L. Lincer, and R.T. Reynolds. 1973. Organochlorines, heavy metals, and the biology of North American Accipiters. BioScience 23:300-305.

Stephen, B.J., J.D. Garlich, and F.E. Guthrie. 1970. Effect of DDT on induction of microsomal enzymes and deposition of calcium in the domestic chicken. Bull. Environ. Contam. Toxicol. 5:569-576.

Temple, S.A. 1972. Chlorinated hydrocarbons residues and reproductive success in eastern North American Merlins. Condor 74: 105-106.

Torda, C., and H. Wolff. 1949. Effect of convulsant and anticonvulsant agents on the activities of carbonic anhydrase. J. Pharmacol. Exp. Ther. 95:444-447.

SYNTHESIS OF THE SYMPOSIUM: ENVIRONMENTAL DYNAMICS OF PESTICIDES

S.M. Lambert

Shell Chemical Company

Houston, Texas 77001

I have been asked to provide for you at this time what has been termed a "Synthesis of the Symposium". Should you ask the question as to what properly constitutes a synthesis, by way of an answer I can only tell you what my approach will be here this morning. Recall, if you will, in the opening paper that Dr. Freed referred to his presentation as an overview. In the same context, mine might be described as a post-posterior view, i.e., one looking back from the end.

I will not attempt to discuss the substantive portion of any of the technical papers; this would be presumptious as the speakers have done an excellent job of presenting their own work, to which I could add nothing of significance and especially in so limited a time. With this in mind, perhaps the most appropriate way to approach the synthesis would be to confine my remarks to some general aspects including impressions I have gained during the symposium. If you will bear with me as I attempt to decipher my very cryptic and hastily scribbled notes, we may, hopefully, develop a synthesis.

The best place to start is at the beginning with the opening paper by Drs. Freed and Haque. Taking some liberties with Dr. Freed's commentary, we heard in essence that the subject matter which constitutes the symposium would focus upon dispersion of chemicals in the environment. Specifically, it would deal with the phenomenon of chemical transport and involve some major questions arising therefrom which relate to the behavior and fate of chemicals under examination.

The subject matter in effect poses for us four principal questions; the first having to do with mechanisms, the second having to do with the flux itself, the third concerning physico-chemical behavior, and the last concerning the biological effect or biological consequence of dispersion. Undergirding each question is yet another question which is both of fundamental importance and common to all; that of the analytical methodology used to observe and measure the phenomena under examination. All of the above questions taken in the broad holistic view of world environmental dynamics are, in effect, posed within the all important framework of production of man's food and protection of man's health.

During the course of the symposium, there have been many highly stylized references and/or slides describing the environment as being made up of four interrelated spheres through which all chemicals must travel at one time or another. There is the atmosphere, the hydrosphere, and the lithosphere, and overlaying all three spheres is what has been referred to as the biosphere.

As a consequence of this very real classification of our environment, there exists the very real requirement of having a broad spectrum of scientific disciplines involved in studies designed to answer the diverse and complex set of questions set before us. The importance of the multi-disciplinary approach to studies in environmental dynamics has certainly been in evidence here. We have heard from chemists (all types - physical, organic, analytical, biological, etc.). We have heard from biologists, entomologists, plant physiologists, and microbiologists. We have heard from chemical engineers, agronomists, marine biologists, physicists, and physicians. This is only a partial listing of the multitude of scientific disciplines necessarily involved in these studies.

I would like now to look at the balance of information presented at the symposium. Granting me some editorial license, I should like to examine this balance in two dimensions. First, on a technical or scientific plane, we have traversed a spectrum ranging from "the primarily theoretical" to that having a very practical, "real world orientation". Second, recalling the four environmental spheres, I have tried to keep a scorecard on the balance of subject matter dedicated to each sphere. On the basis of having both primary and/or secondary emphasis, my scorecard reads as follows: six papers placed a primary and/or secondary emphasis on the atmosphere; five papers on the hydrosphere; seven on the lithosphere; and six on the biosphere. All in all presenting us with a very good overall subject balance.

After hearing the papers presented over the past day and a half, I thought it might be instructive to construct in narrative form what one might perceive as being the major purpose of this symposium. This I would state as follows:

"To illuminate principal forces at work and the complex
relationships that exist among them which should be useful in
providing quantitative, realistic descriptions of how chemicals
disperse in our environment and of the real and potential biological
consequences attendant with this dispersion."

A point worthy of mention at this time concerns an appeal made
in one form or another by a number of the symposium speakers. The
appeal takes the form of a request for more comprehensive descriptions of what we, in fact, observe or what we believe to be
occurring as chemicals under examination move and are dispersed
through our environment. These authors apparently see the need for
taking a broader view, i.e., one which encompasses most, if not
all, relevant aspects of the problem at hand (as opposed to a more
narrow view). Certainly, the broad, more comprehensive viewpoint
is an aid to better understanding. It does, of course, suffer the
disadvantage from an experimental point of adding greater degrees
of complexity. Nevertheless, I believe most of us would agree that
to the extent that scientists in this field can increase the
dimensionality of their experimentation, this would be desirable.

As an aid to providing a more comprehensive view of the dispersion of chemicals, I should like to direct your attention to
the figure, an attempt I have made to describe conceptually and
schematically principal elements which should, at one time or
another, be taken into consideration. Looking first at elements
of flows, one sees the primary source of product. Our concern begins
with production and insertion of the chemical into the environment.
As the chemical is dispersed, our concern turns to the chemical
flux in terms of initial receptors and secondary sources. And
finally, we must consider the chemical sinks and/or ultimate fate.

As for continuity of flow, beginning with production we
proceed through various levels of dispersion coupled to various
levels of use and ultimately through what is termed major use
trails. In this manner, the scheme does consider the evolution
of end-use products. Concomitant with the continuity of flow is
the very important consideration of new species identification.
This is denoted by a transformation continuum which attempts to
describe primarily chemical and/or biological transformations, but
to some degree also considers changes of a purely physical nature,
i.e., changes of state.

Finally, we concern ourselves with the impact on the environment through an assessment of real and potential hazard. This
takes the form of an analytical continuum of environmental exposure
through all phases of dispersion and use.

Now, as to the symposium itself, I think you'll agree that it
has been both interesting and informative. We've been provided with

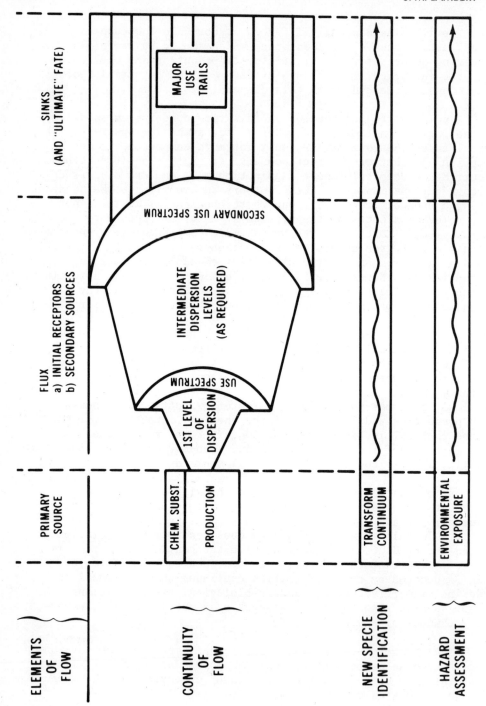

information on how pesticides behave; specifically, how they move in our environment and the effects or consequences of these actions. We've heard that these materials obey strict physical laws as they disperse and that modeling this phenomena is important not only as a productive tool, but as a way to help us gain some understanding of what is, in fact, occurring, especially as pertains to forces which govern movement of chemicals.

With regard to analytical methodology, many authors have stressed the importance of tools of measurement which are required not only to validate and confirm, but also to assist in construction of model descriptions.

Finally, I should like to share with you four distinct and important messages I have discerned in the course of this symposium. See if you agree with these impressions.

The first involves a message given from those experienced workers to the less experienced who may be planning work or have work in progress.

"In our experience, we find these factors (those we have discussed) to be of major importance or consequence in understanding and describing the phenomena we observe. Please examine these factors at least as 'points of departure' in your own work."

The second: "In most areas in the field of dynamics of pesticides our understanding is very rudimentary and quite incomplete. Only in a very few cases is our knowledge and understanding well developed."

The third: "We need more good work in this area; a must if we are eventually to gain better insight and greater understanding of the phenomena we observe".

The fourth: "The subject matter is indeed complex - that is obvious - but, and a very important but, we are making headway - a significant headway in understanding. Let's see to it that this progress continues - IT IS OF VITAL IMPORTANCE TO MAN, HIS FOOD PRODUCTION, AND HIS HEALTH".

In closing, Drs. Freed and Haque deserve our sincere appreciation for putting together what I'm sure you'll agree has been a truly first-class symposium. Together with the symposium speakers, let's give them all a round of applause for a very fine job indeed.

CONTRIBUTORS

S.H. Ashrafi
    Department of Biological Sciences
    University of Illinois at Chicago Circle
    Chicago, Illinois 60680

E.J. Baum
    Oregon Graduate Center
    Beaverton, Oregon 97005

John W. Brewer
    Department of Mechanical Engineering
    University of California
    Davis, California 95616

M.M. Cliath
    USDA, Agricultural Research Service
    University of California
    Riverside, California 92502

W. Brian Crews
    Department of Mechanical Engineering
    University of California
    Davis, California 95616

D.G. Crosby
    Department of Environmental Toxicology
    University of California
    Davis, California 95616

J.E. Davies
    Department of Epidemiology and Public Health
    University of Miami School of Medicine
    Miami, Florida 33152

Henry F. Enos
    Division of Equipment and Techniques
    U.S. Environmental Protection Agency
    Washington, D.C. 20460

Raymond A. Fleck
     Food Protection and Toxicology Center
     University of California
     Davis, California 95616

V.H. Freed
     Department of Agricultural Chemistry and
     Environmental Health Sciences Center
     Oregon State University
     Corvallis, Oregon 97331

M.L. Gassman
     Department of Biological Sciences
     University of Illinois at Chicago Circle
     Chicago, Illinois 60680

C.A.I. Goring
     Ag-Organics Department
     Dow Chemical, U.S.A.
     Midland, Michigan 48640

John W. Hamaker
     Ag-Organics Research Department
     Dow Chemical, U.S.A.
     Walnut Creek, California 94598

R. Haque
     Department of Agricultural Chemistry and
     Environmental Health Sciences Center
     Oregon State University
     Corvallis, Oregon 97331

Rolf Hartung
     Environmental and Industrial Health
     School of Public Health
     University of Michigan
     Ann Arbor, Michigan 48104

John P. Hassett
     Institute for Environmental Sciences
     University of Texas-Dallas
     Dallas, Texas 75080

Dennis P.H. Hsieh
     Food Protection and Toxicology Center
     University of California
     Davis, California 95616

Haji M. Jameel
    Food Protection and Toxicology Center
    University of California
    Davis, California 95616

Eugene E. Kenaga
    Health and Environmental Research Department
    Dow Chemical, U.S.A.
    Midland, Michigan 48640

M.A.Q. Khan
    Department of Biological Sciences
    University of Illinois at Chicago Circle
    Chicago, Illinois 60680

Wendell W. Kilgore
    Food Protection and Toxicology Center
    University of California
    Davis, California 95616

S.M. Lambert
    Shell Chemical Company
    Houston, Texas 77001

D.A. Laskowski
    Ag-Organics Department
    Dow Chemical, U.S.A.
    Midland, Michigan 48640

G. Fred Lee
    Institute for Environmental Sciences
    University of Texas-Dallas
    Dallas, Texas 75080

Ming Y. Li
    Food Protection and Toxicology Center
    University of California
    Davis, California 95616

R.W. Meikle
    Ag-Organics Department
    Dow Chemical, U.S.A.
    Walnut Creek, California 94598

K.W. Moilanen
    Department of Environmental Toxicology
    University of California
    Davis, California 95616

R.D. O'Brien
    Section of Neurobiology and Behavior
    Cornell University
    Ithaca, New York 14850

Ruth R. Painter
    Food Protection and Toxicology Center
    University of California
    Davis, California 95616

David B. Peakall
    Section of Ecology and Systematics
    Cornell University
    Ithaca, New York 14850

Timothy J. Petersen
    Department of Mechanical Engineering
    University of California
    Davis, California 95616

R.L. Pitter
    Oregon Graduate Center
    Beaverton, Oregon 97005

S.A. Poznanski
    Department of Epidemiology and Public Health
    University of Miami School of Medicine
    Miami, Florida 33152

James N. Seiber
    Department of Environmental Toxicology
    University of California
    Davis, California 95616

Talaat M. Shafik
    Pesticides and Toxic Substances Effects Laboratory
    U.S. Environmental Protection Agency
    Research Triangle Park, North Carolina 27711

R.F. Smith
    Department of Entomological Sciences
    University of California
    Berkeley, California 94720

C.J. Soderquist
    Department of Environmental Toxicology
    University of California
    Davis, California 95616

W.F. Spencer
    USDA, Agricultural Research Service
    University of California
    Riverside, California 92502

A.S. Wong
    Department of Environmental Toxicology
    University of California
    Davis, California 95616

James E. Woodrow
    Department of Environmental Toxicology
    University of California
    Davis, California 95616